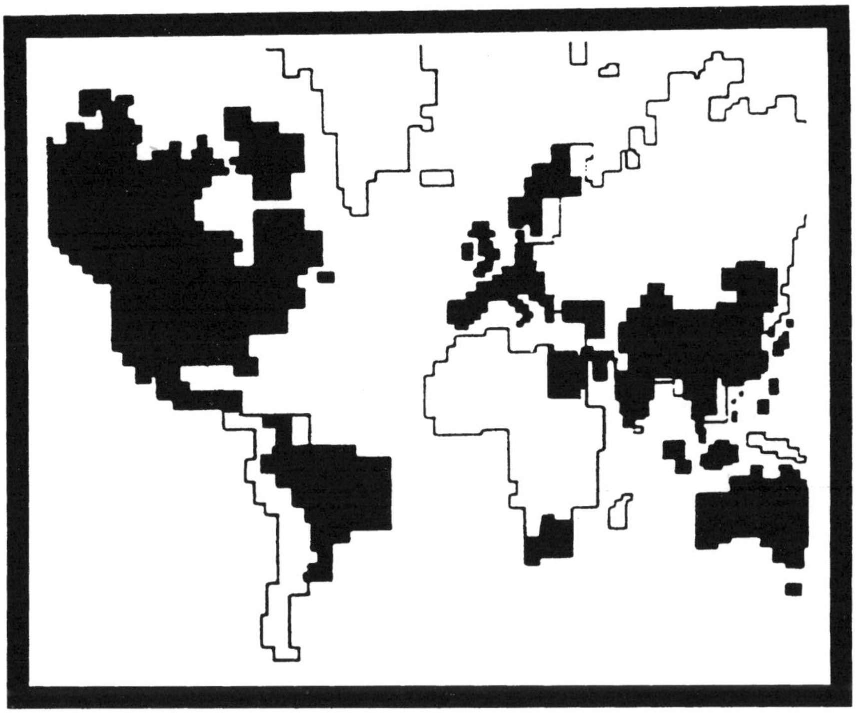

Yearbook of World Electronics Data 1994

Vol. 3: Emerging Countries & World Summary

MARKET PROSPECTS TO 1997

ELSEVIER
ADVANCED
TECHNOLOGY

YEARBOOK OF WORLD ELECTRONICS DATA 1994 VOLUME 3 - EMERGING COUNTRIES & WORLD SUMMARY

Research Consultant: Kenneth F. Wilson
Publishing Manager: Andrew Fletcher

First Edition Published 1990
THIRD EDITION PUBLISHED: July 1994
ISSN: 0954-0180
ISBN: 1 85617 202 3

British Library Cataloguing in Publication Data
A CIP Catalogue record for this book is available from the British Library

Elsevier Advanced Technology
The Boulevard
Langford Lane
Kidlington
Oxford OX5 1GB, United Kingdom
Tel: (+44)(0)865 843848/843000
Fax: (+44)(0)865 843971

LIST OF CONTENTS

PREFACE

This volume of the Yearbook is the third edition of the 'Emerging Countries'. Included in its coverage of 10 countries is China, which is becoming increasingly important on the world electronics industry stage. Also included is Mexico which is an important producer of electronic goods, mainly for the US. Greece and Portugal are included to complete the coverage of the Economic Union countries in the Yearbook programme.

Research for the Yearbook now covers 48 countries with the results published in four volumes. By covering most of the industrialised world we are able to make country comparisons thus aiding estimation in those areas where data is unavailable or verification of statistics is unreliable.

In 1993 we updated our East Europe volume reporting on the changes occurring since the break-up of the communist block. At the end of 1993 we published the twenty first edition of the West Europe volume and early in 1994 the eleventh edition of the America, Japan & Asia Pacific volume. These were prepared as the world is slowly emerging from recession and, together with this volume, our analysis of the very latest available data gives us an unique insight into recent trends in the electronics industry and enables a realistic assessment of future prospects to be made.

Towards the end of 1994 we will be publishing the twenty second edition of the West Europe volume. Research for this is likely to be complicated by the changes in the collection of intra-EU trade, this now being done on a sampling basis. Inevitably this will complicate and delay statistical analysis for all researchers and an overall 'world look' becomes even more valuable. However, subscribers can be assured that we will continue to produce the Yearbook and make it as up to date and as comparable with past data as possible.

Kenneth F. Wilson
Research Consultant
July 1994

1 INTRODUCTION

The Yearbook series presents a review of the world electronics industries and markets with detailed estimates and forecasts of the size of the production and markets for electronic equipment and components in each country.

The major 30 countries of the world are surveyed annually and this is supplemented by a biannual survey of 10 'emerging countries' and by a biannual survey of the East European countries.

The 30 major nations are:

Western Europe — Austria, Belgium, Denmark, Finland, France, Germany, Ireland, Italy, Netherlands, Norway, Spain, Sweden, Switzerland, UK.

America, Japan, Asia Pacific — Australia, Brazil, Canada, Hong Kong, India, Indonesia, Israel, Japan, Malaysia, Philippines, Singapore, South Africa, South Korea, Taiwan, Thailand, USA.

This volume of the Yearbook covers the emerging countries in detail — China, Egypt, Greece, Mexico, New Zealand, Portugal, Puerto Rico, Saudi Arabia, Turkey, Venezuela.

GENERAL YEARBOOK METHODOLOGY

The overall methodology employed in the compilation of the Yearbooks is to analyze trade and production statistics in each of the countries. The aim is to cover all electronic products, and electronic components used in those products, and these are grouped into 10 major categories which are common across all countries. The degree of detail within each category varies for each county depending upon the importance of the country and the availability of statistical sources.

Trade statistics are analyzed in detail, over 350 separate categories being employed. For some of the emerging countries, where trade statistics from a country are limited or unreliable, as in China, greater emphasis has been placed on trade statistics of countries trading with that country. Production statistics are collected from Government and Manufacturer's Association sources where these are available. Extensive use is also made of research reports, company reports, news items and work by other consultants to supplement and cross check the official and semi-official sources.

From this work the 'base year' estimates are built up, this being 1992 for the 1994 edition of the Yearbook.

Production is then forecast for the next 2 years and markets for the next 5 years. These forecasts are made in **real terms using constant exchange rates and excluding inflation.** Forecasts are based on historical performance of each product (of which we have extensive knowledge), the general economic outlook and the major growth influencing factors. Greater detail of this aspect of the methodology is given in the 'guide to the interpretation of the statistics' in Appendix 5.3.

Each new edition of the Yearbook is **fully revised** and **updated** with all sources being reanalyzed and therefore it must be emphasised that the constantly changing electronics industry requires a continuing programme of revisions to the market and production data. This methodology does not, therefore, facilitate year-on-year comparisons of products between different editions, but historical figures on a comparable basis are available on PC Diskette for the major countries only.

2 WORLD SUMMARY DATA

2.1 ECONOMIC OVERVIEW

Growth in the major economies of the world, as represented by the OECD, was only 1.0% in 1991 and 1.6% in 1992 and remained low in 1993 at 1.1%. Growth will accelerate 2.1% in 1994 and 2.5% in 1995, held back by weak economic activity in Japan and Europe but helped as the US recovery continues. The annual average during 1994-98 will be 2.5%. Growth will be dampened by widespread and deep-seated structural impediments, including job insecurity, high indebtedness, large government deficits and weak real estate markets. High real interest rates currently remain a major obstacle to recovery in Europe.

The recovery in the United States, which began as long ago as early 1991, was sluggish and patchy but is now gathering momentum. After gathering pace late in 1992 it slowed during the first half of 1993 but achieved a slightly stronger performance in the second half of the year. In annual average terms GDP expanded by 2.6% in 1992, 3.0% in 1993 and 3.5% is forecast for 1994 before slowing again in 1995. This is the net result of several conflicting forces on the key sector of the economy which is consumer demand. The main positive stimulus comes from the low level of interest rates. Consumer spending is, however, being held back by a lack of confidence, which may reflect concern about the still high level of consumer debt, but also reflects fear about job security and employment prospects. However, over the 5 year period 1994-1998 the US will out-perform both Japan and Western Europe with GDP growth averaging 2.8% per annum.

Japan's economy is suffering from a fully-fledged recession, real GDP for 1993 was around 0% compared with 1.1% in 1992 and 4.1% in 1991. Three recent fiscal stimulation packages totalling ¥30 trillion, the prospect of income tax cuts and a record low Discount rate have so far only slightly revived a private sector which is suffering poor corporate profits, a strong yen, weak personal income growth, rising job insecurity and high indebtedness. Growth in 1994 is forecast to be only 0.3% and to grow by 2.0% in 1995.

Rising unemployment, weak consumer and business confidence, high real interest rates, weak export markets and large government deficits are widespread problems throughout Europe. Spain, Italy, the UK and Ireland have benefited from currency devaluation. Weakness is primarily centred in manufacturing, although in France at least the recession is increasingly spreading to the service sector. With EU unemployment set to continue rising from the current level of 17.5 million towards 20 million in 1994, recovery will develop only slowly. EU GDP has fallen steadily since 1990, being 1.5% in 1991 and 1.0% in 1992. In 1993 it fell 0.4%, followed by a modest 1.5% growth forecast for 1994. Over the medium term EU GDP growth will be limited to an average of 2.2% per annum during 1994-98, below the US and Japan.

Most of the Asia-Pacific countries continue to grow strongly in the face of the poor performance of the major countries and depressed world trade. Excluding Japan, growth in the Asia-Pacific region was 6.3% in 1992, 6.9% in 1993 and is forecast to fall only slightly in 1994 and 1995. Malaysia is particularly strong, closely followed by Thailand with only the Philippines being relatively depressed.

Inflation will remain subdued reflecting wage moderation amid job insecurity, increased global competition and weak oil prices. OECD consumer prices are forecast to rise by only 2.5% in 1994 and 3.1% in 1995.

2.2 ELECTRONICS MARKET OVERVIEW OF THE MAJOR COUNTRIES

The market for electronics equipment and components in the 30 major countries of the world in West Europe, America, Japan and Asia Pacific was US$683 billion in 1992. This was an increase of 0.1% in current dollar terms. However, if exchange rate variations and inflation are excluded the market decreased by 5.4%.

In 1993 the electronics market started to recover, although, with the recession still evident in some countries, the overall real growth was only 0.2%. It will be 1994 before any significant world growth is apparent, forecast at 2.3% for the year. The rate of growth will increase, to 3% in 1995 and 3.3% in 1996, but in 1997 many countries will experience a slowing with growth forecast at around 3%. These rates of growth, low by comparison to those seen in the nineteen eighties, are due to depressed military expenditure, the effects of low cost production in countries such at China, Vietnam, North Korea and Russia and low rates of economic growth in many countries. It should be noted that throughout the Yearbook series all figures, with the exception of 1991, are expressed in 1992 constant exchange rate and values and therefore growth rates are real, excluding inflation. It should also be noted that production and market estimates and forecasts are based on factory sales.

Following a real decline of 5% in 1991, the US electronics market grew by 1.5% in 1992 compared with a forecast growth of 1.2% made in the 1993 Yearbook. The main growth areas were electronic data processing at 5.8%, consumer products at 4.4%, active components at 7% and telecommunications at 6.5%. The communications and radar sector, which includes military communications, declined by 5%, with radio communications including mobile telephones growing by 3.6%. In 1993 real growth is estimated to have been 4% with integrated circuits growing by 17%. Growth in 1994 is forecast to be similar in most sectors except integrated circuits, which will only increase by 7%, due to stabilising of prices. In 1995 growth is forecast to slow to only 1.9% and then to grow at a subdued level of around 2.5% per annum for the next two years.

The Japanese market was disastrous in 1992 with a staggering real fall of 17.6%. The worst sector was consumer, with a fall of over 21%, colour television factory sales declining from 8.7 million units to 7.1 million units. Most sectors exhibited similar reductions except for medical and office equipment. Further large declines occurred in 1993, notably in electronic data processing, 13% down, and consumer, 15% down, with the overall drop being estimated as 8.3%. The recession is forecast to continue in 1994 with a further fall in the market of 2.2%. The stimulating measures taken by the government will eventually have an effect and in 1995 growth will resume. In 1996 and 1997 Japan is expected to have the highest growth for electronic products in the industrialised world approaching 6% p.a., although the market size in 1997 in yen terms will still only be around that of 1988.

Production in Japan declined by a real 12.7% in 1992 and by an estimated 6.1% in 1993, less than the drop in the domestic market. However, in the longer term the move to overseas production continues with the number of foreign plants increasing. It is likely that growth in

Japanese production in the mid-to late-1990s will be at the world level for industrialised countries of 3-4%, lower than growth in the domestic market.

The West Europe market fell by 7.2% in 1992 and by a further 1.1% in 1993. However, there were large differences between the countries with Germany declining by 11.5% in 1992 and 2.3% in 1993 whilst in the UK the decline was only 1.4% in 1992 with growth of 2.7% in 1993. In 1994 all countries of West Europe should return to growth, although the level will be a modest 2% in total. In 1995 and 1996 it is forecast that growth will exceed 3% but in 1997 there will be a slowing to around 2%. Over this period growth in Germany is forecast to increase to 4.5% in 1996 while the UK will remain at around 2.7% p.a. from 1992 to 1996, reducing to 1.7% in 1997.

Growth is resuming in the Brazilian economy and with it the electronics market. Following a decline of 2.4% in 1992 growth for electronics products is forecast at 5.5% in 1993, increasing to 7.2% in 1994. After this the growth rate is expected to fall slightly but to be still buoyant at 4.6% in 1997. Relaxation of trade tariffs are expected to increase exports with the result that production of electronic goods is forecast to increase by 8% in 1993 and 11% in 1994.

The Asia Pacific region, excluding Japan, remains buoyant with growth in production exceeding 6% per annum in 1992 to 1994. The countries with the highest growths are Indonesia, Malaysia and Thailand with an average growth of 17.5% p.a. South Korea, however, has followed the pattern of the industrialised world with a market decline of 6.3% in 1992 and only minimal production growth of 1.2%. By 1994 growth in both its domestic market and its production should return to a level of around 4% p.a. Hong Kong continues its progression to a service centre for Chinese production with high market growth and low production growth.

SUMMARY OF MAJOR COUNTRIES PRODUCTION 1991

Table 2.3.1

Country	$M EDP	$M Office Equipment	$M Control & Instr	$M Medical & Industrial	$M Comms & Radar	$M Telecomm–unications	$M Consumer	$M Components	$M TOTAL
Australia	688	28	234	91	493	820	135	261	**2752**
Austria	348	49	336	98	49	340	1174	994	**3387**
Belgium	1701	24	503	167	313	1210	717	760	**5395**
Brazil	4843	318	620	225	740	1280	2270	2433	**12729**
Canada	2868	119	917	434	1090	2234	339	624	**8626**
Denmark	150	22	399	218	208	121	142	467	**1726**
Finland	657	6	394	168	265	405	181	434	**2510**
France	7303	587	2963	835	7553	4649	1808	4571	**30269**
Germany	10117	1301	9828	2948	3328	7350	4140	11067	**50079**
Hong Kong	2145	306	109	96	403	512	2743	2026	**8340**
India	455	90	261	59	519	544	1115	735	**3778**
Indonesia	165	37	50	42	176	224	647	312	**1653**
Ireland	3218	35	298	44	131	302	13	944	**4984**
Israel	330	7	170	330	523	380	170	448	**2358**
Italy	7035	450	2825	779	2659	4953	1041	2754	**22497**
Japan	59563	5400	7519	5452	10778	15519	36519	66652	**207400**
Malaysia	771	89	97	65	441	665	2814	4147	**9091**
Netherlands	2235	842	1461	893	610	1010	94	1274	**8419**
Norway	350	0	248	69	229	414	7	119	**1436**
Philippines	135	12	30	19	210	132	156	1445	**2139**
Singapore	7977	302	261	71	373	242	2038	5445	**16710**
South Africa	192	7	45	56	170	518	215	98	**1301**
South Korea	3499	253	202	186	766	1884	6697	11959	**25446**
Spain	1640	32	274	233	506	2649	1207	837	**7379**
Sweden	967	38	995	320	1247	1940	103	836	**6446**
Switzerland	540	109	1891	469	255	571	2304	848	**6987**
Taiwan	5938	239	83	217	708	1368	1713	5513	**15779**
Thailand	1908	119	57	31	184	365	1149	1590	**5403**
UK	8611	730	4202	1140	4274	2474	2500	4793	**28723**
USA	47965	5282	26673	9588	50562	16259	6060	44024	**206413**
TOTAL	**184314**	**16833**	**63946**	**25342**	**89764**	**71335**	**80211**	**178410**	**710155**

Notes: 1991 and 1992 are current figures at current exchange rates
1993 & 1994 are forecasts at 1992 constant values & exchange rates (ie. inflation is not included)
West Europe figures are taken from Volume 1 of the 1994 edition of the Yearbook
US, Japan & Asia Pacific figures are taken from Volume 2 of the 1994 edition of the Yearbook

SUMMARY OF MAJOR COUNTRIES PRODUCTION 1992

Table 2.3.2

Country	$M EDP	$M Office Equipment	$M Control & Instr	$M Medical & Industrial	$M Comms & Radar	$M Telecomm–unications	$M Consumer	$M Components	$M TOTAL
Australia	746	22	272	94	464	779	142	258	**2778**
Austria	444	48	341	92	53	460	1078	938	**3453**
Belgium	1701	29	493	183	362	1154	701	725	**5347**
Brazil	4614	329	630	238	760	1200	2331	2607	**12709**
Canada	2835	100	950	459	1140	2066	350	520	**8420**
Denmark	169	19	436	252	162	126	152	541	**1857**
Finland	705	6	348	171	415	410	136	376	**2567**
France	7404	551	3002	898	7696	5070	1837	4844	**31302**
Germany	8181	1290	10265	3252	3440	7985	3288	10779	**48480**
Hong Kong	2261	287	89	121	517	459	2613	1973	**8320**
India	453	89	232	65	501	610	1098	782	**3830**
Indonesia	275	49	55	44	167	220	904	449	**2163**
Ireland	3729	39	244	47	164	281	15	1163	**5683**
Israel	422	7	200	359	575	447	122	596	**2728**
Italy	6901	406	2980	803	2852	4898	954	2698	**22490**
Japan	56598	5559	6898	5543	11142	14976	31638	63614	**195969**
Malaysia	1686	108	121	78	568	778	3451	5339	**12128**
Netherlands	2317	903	1573	880	625	1080	128	1274	**8781**
Norway	332	0	259	75	225	364	7	112	**1373**
Philippines	125	12	29	20	200	120	159	1440	**2105**
Singapore	10123	332	290	79	485	264	2484	6196	**20252**
South Africa	207	7	53	59	175	500	216	91	**1308**
South Korea	3647	258	211	192	883	1798	6363	13065	**26417**
Spain	1544	34	248	237	433	1959	1169	806	**6431**
Sweden	876	33	1042	306	1344	1912	57	828	**6399**
Switzerland	567	97	1883	474	298	557	2489	898	**7262**
Taiwan	7865	203	88	242	732	1249	1542	6080	**18001**
Thailand	2313	153	87	45	239	429	1456	1984	**6705**
UK	8293	686	4011	1281	4604	2093	1847	4886	**27700**
USA	47942	5117	26711	9960	49391	18089	6254	46304	**209768**
TOTAL	**185274**	**16772**	**64041**	**26547**	**90610**	**72333**	**74981**	**182167**	**712725**

Notes: 1991 and 1992 are current figures at current exchange rates
1993 & 1994 are forecasts at 1992 constant values & exchange rates (ie. inflation is not included)
West Europe figures are taken from Volume 1 of the 1994 edition of the Yearbook
US, Japan & Asia Pacific figures are taken from Volume 2 of the 1994 edition of the Yearbook

SUMMARY OF MAJOR COUNTRIES PRODUCTION 1993

Table 2.3.3

Country	$M EDP	$M Office Equipment	$M Control & Instr	$M Medical & Industrial	$M Comms & Radar	$M Telecomm–unications	$M Consumer	$M Components	$M TOTAL
Australia	799	23	291	104	473	772	142	274	**2877**
Austria	488	46	313	92	54	437	1048	919	**3396**
Belgium	1667	29	478	188	377	1096	738	714	**5286**
Brazil	5075	345	662	262	874	1236	2448	2842	**13744**
Canada	2976	103	1007	511	1255	2087	370	544	**8853**
Denmark	166	18	423	243	160	122	148	514	**1795**
Finland	670	6	338	171	456	389	150	379	**2559**
France	7478	529	2852	898	7773	4918	1829	4960	**31235**
Germany	8099	1225	9855	3031	3578	7905	3399	10686	**47779**
Hong Kong	2306	273	92	126	568	468	2586	1958	**8377**
India	476	93	220	71	531	671	1193	879	**4134**
Indonesia	371	54	58	48	184	231	1055	604	**2604**
Ireland	3915	40	249	49	174	267	15	1212	**5922**
Israel	464	7	220	381	604	492	110	680	**2958**
Italy	6763	402	3009	819	2966	4604	1039	2735	**22337**
Japan	50939	5003	7036	5556	11365	14078	26734	63213	**183923**
Malaysia	2360	117	130	84	653	817	3826	6151	**14138**
Netherlands	2248	867	1542	861	644	1015	131	1237	**8544**
Norway	335	0	257	76	229	345	7	110	**1360**
Philippines	135	13	31	23	210	122	167	1514	**2214**
Singapore	10933	339	301	83	533	267	2487	6516	**21458**
South Africa	211	7	59	60	173	460	216	84	**1271**
South Korea	3756	263	215	202	971	1762	6203	13855	**27228**
Spain	1497	35	255	236	446	1665	1145	801	**6080**
Sweden	832	32	1001	307	1384	1759	59	826	**6201**
Switzerland	573	99	1902	483	313	540	2538	927	**7376**
Taiwan	8258	193	90	247	754	1299	1272	6508	**18621**
Thailand	2659	176	113	57	298	471	1714	2434	**7922**
UK	8459	700	4051	1340	4696	1967	2054	5117	**28383**
USA	49380	5219	27245	10417	48403	18451	6384	49628	**215128**
TOTAL	**184290**	**16255**	**64293**	**27026**	**91099**	**70715**	**71208**	**188819**	**713704**

Notes: 1991 and 1992 are current figures at current exchange rates
1993 & 1994 are forecasts at 1992 constant values & exchange rates (ie. inflation is not included)
West Europe figures are taken from Volume 1 of the 1994 edition of the Yearbook
US, Japan & Asia Pacific figures are taken from Volume 2 of the 1994 edition of the Yearbook

SUMMARY OF MAJOR COUNTRIES PRODUCTION 1994

Table 2.3.4

Country	$M EDP	$M Office Equipment	$M Control & Instr	$M Medical & Industrial	$M Comms & Radar	$M Telecomm–unications	$M Consumer	$M Components	$M TOTAL
Australia	846	24	314	112	492	764	146	288	**2987**
Austria	512	47	320	96	56	445	1021	947	**3445**
Belgium	1701	30	488	192	384	1074	764	731	**5363**
Brazil	5837	373	728	298	1005	1298	2621	3185	**15344**
Canada	3096	106	1058	562	1380	2108	391	566	**9265**
Denmark	171	19	436	248	165	121	151	526	**1836**
Finland	690	6	341	176	466	381	157	392	**2609**
France	7628	539	2880	922	8006	5016	1869	5195	**32055**
Germany	8342	1201	9756	2987	3685	8064	3463	11064	**48562**
Hong Kong	2375	267	96	131	625	477	2579	2025	**8577**
India	523	96	209	79	574	738	1315	1010	**4545**
Indonesia	483	57	61	52	211	247	1220	772	**3102**
Ireland	4111	41	256	50	183	254	15	1255	**6166**
Israel	501	8	242	406	640	531	104	753	**3185**
Italy	6966	410	3100	848	3025	4512	1069	2814	**22743**
Japan	48392	4753	7458	5823	12046	13937	25848	65838	**184095**
Malaysia	3068	124	141	91	751	858	4098	6904	**16035**
Netherlands	2315	859	1557	871	657	974	134	1248	**8615**
Norway	348	0	267	79	236	349	7	114	**1401**
Philippines	149	13	33	25	227	126	179	1662	**2415**
Singapore	12026	349	316	88	576	272	2457	6773	**22857**
South Africa	222	7	66	63	180	446	225	85	**1295**
South Korea	3907	268	222	215	1068	1780	6261	14624	**28345**
Spain	1542	35	263	241	468	1582	1178	814	**6125**
Sweden	866	33	1021	314	1439	1707	63	849	**6292**
Switzerland	590	101	1940	501	328	529	2615	961	**7564**
Taiwan	8754	183	92	254	784	1325	1092	6862	**19345**
Thailand	3058	202	158	71	388	542	2096	2976	**9491**
UK	8628	714	4091	1389	4789	1889	2383	5336	**29219**
USA	51849	5324	28063	11084	47919	18820	6560	51866	**221485**
TOTAL	**189495**	**16188**	**65971**	**28267**	**92756**	**71166**	**72083**	**198436**	**734362**

Notes: 1991 and 1992 are current figures at current exchange rates
1993 & 1994 are forecasts at 1992 constant values & exchange rates (ie. inflation is not included)
West Europe figures are taken from Volume 1 of the 1994 edition of the Yearbook
US, Japan & Asia Pacific figures are taken from Volume 2 of the 1994 edition of the Yearbook

SUMMARY OF MAJOR COUNTRIES MARKETS 1991

Table 2.4.1

Country	$M EDP	$M Office Equipment	$M Control & Instr	$M Medical & Industrial	$M Comms & Radar	$M Telecomm–unications	$M Consumer	$M Components	$M TOTAL
Australia	2469	174	688	211	758	956	821	837	**6913**
Austria	1454	121	525	171	186	355	669	924	**4405**
Belgium	2820	158	749	218	379	1082	672	1317	**7395**
Brazil	5050	342	927	353	866	1298	2114	3140	**14090**
Canada	5147	422	1626	647	1305	2000	1888	2582	**15617**
Denmark	938	76	139	78	123	226	339	578	**2495**
Finland	1079	61	376	93	206	332	293	697	**3137**
France	9944	821	3561	1033	5490	3976	3451	6865	**35142**
Germany	16280	1827	7805	2050	3107	6622	8014	11930	**57636**
Hong Kong	1093	61	122	78	179	372	708	3362	**5975**
India	519	100	359	131	540	578	1074	929	**4231**
Indonesia	289	62	248	115	365	374	577	585	**2615**
Ireland	1106	60	218	77	116	240	137	1011	**2966**
Israel	598	31	224	225	420	230	188	570	**2486**
Italy	8332	610	3709	817	2590	5603	3637	4921	**30219**
Japan	44163	2022	6111	3800	7933	11978	15785	48896	**140689**
Malaysia	472	45	429	77	349	450	378	2617	**4817**
Netherlands	5351	457	1560	585	847	1230	1426	1836	**13293**
Norway	944	69	342	105	253	425	240	373	**2752**
Philippines	157	21	64	22	117	220	135	357	**1093**
Singapore	2582	168	584	82	275	381	689	4018	**8778**
South Africa	842	142	279	126	233	620	392	411	**3045**
South Korea	2475	219	1484	346	715	1507	2281	6782	**15809**
Spain	3578	358	919	529	688	2941	2506	2139	**13657**
Sweden	2076	144	888	191	1032	853	741	1536	**7461**
Switzerland	2441	312	1155	294	393	683	778	1008	**7062**
Taiwan	924	84	673	193	384	1007	861	4676	**8803**
Thailand	787	46	312	102	298	227	457	1460	**3688**
UK	10288	800	3939	1104	3956	2711	3432	7261	**33489**
USA	46566	6611	22460	9154	49812	18028	19204	50197	**222032**
TOTAL	**180765**	**16424**	**62474**	**23005**	**83916**	**67505**	**73885**	**173816**	**681789**

Notes: 1991 and 1992 are current figures at current exchange rates
1993 to 1997 are forecasts at 1992 constant values & exchange rates (ie. inflation is not included)
West Europe figures are taken from Volume 1 of the 1994 edition of the Yearbook
US, Japan & Asia Pacific figures are taken from Volume 2 of the 1994 edition of the Yearbook

SUMMARY OF MAJOR COUNTRIES MARKETS 1992

Table 2.4.2

Country	$M EDP	$M Office Equipment	$M Control & Instr	$M Medical & Industrial	$M Comms & Radar	$M Telecomm–unications	$M Consumer	$M Components	$M TOTAL
Australia	2510	190	699	215	799	946	871	872	**7101**
Austria	1462	127	535	192	206	502	646	963	**4633**
Belgium	2984	153	774	243	432	950	757	1319	**7610**
Brazil	5000	351	952	357	930	1268	2147	3110	**14115**
Canada	4865	432	1626	664	1382	1796	1871	2642	**15279**
Denmark	1016	83	146	88	89	227	345	654	**2647**
Finland	939	48	338	77	182	286	233	759	**2863**
France	10408	807	3568	1155	5783	4285	3356	6720	**36083**
Germany	16197	1710	7758	2279	3281	7131	6803	11972	**57132**
Hong Kong	1339	66	153	95	254	382	743	3472	**6503**
India	530	91	344	131	520	639	1047	982	**4285**
Indonesia	302	56	286	138	440	428	617	715	**2982**
Ireland	1103	59	231	119	122	244	146	1212	**3236**
Israel	752	39	263	235	465	250	219	647	**2870**
Italy	8576	614	3864	844	2786	5413	3184	5079	**30359**
Japan	38598	2244	5118	4079	7874	10992	13433	43157	**125496**
Malaysia	603	53	472	98	432	490	427	3127	**5702**
Netherlands	5786	394	1660	672	853	1207	1383	1970	**13926**
Norway	1006	73	357	111	263	419	268	396	**2893**
Philippines	165	22	66	23	110	210	141	339	**1076**
Singapore	2872	177	669	99	323	417	721	4799	**10077**
South Africa	870	97	302	121	246	632	375	432	**3075**
South Korea	2247	201	1418	317	803	1436	2175	6609	**15206**
Spain	3366	322	828	466	562	2156	2478	1982	**12160**
Sweden	2111	145	947	214	1038	825	633	1607	**7520**
Switzerland	2482	311	1118	324	382	664	682	1003	**6965**
Taiwan	1054	95	755	228	486	860	1032	5821	**10331**
Thailand	1027	56	328	116	383	271	564	1828	**4571**
UK	11219	765	3704	1295	4495	2502	3219	7353	**34551**
USA	50480	6772	22687	9417	48600	19661	20539	53112	**231268**
TOTAL	**181868**	**16551**	**61965**	**24412**	**84520**	**67489**	**71057**	**174652**	**682515**

Notes: 1991 and 1992 are current figures at current exchange rates
1993 to 1997 are forecasts at 1992 constant values & exchange rates (ie. inflation is not included)
West Europe figures are taken from Volume 1 of the 1994 edition of the Yearbook
US, Japan & Asia Pacific figures are taken from Volume 2 of the 1994 edition of the Yearbook

SUMMARY OF MAJOR COUNTRIES MARKETS 1993

Table 2.4.3

Country	$M EDP	$M Office Equipment	$M Control & Instr	$M Medical & Industrial	$M Comms & Radar	$M Telecomm–unications	$M Consumer	$M Components	$M TOTAL
Australia	2636	195	726	234	822	927	915	912	**7368**
Austria	1432	124	524	200	212	492	633	965	**4584**
Belgium	2954	146	750	249	450	902	724	1355	**7529**
Brazil	5250	362	1009	386	1004	1243	2254	3385	**14893**
Canada	4963	441	1692	709	1479	1760	1927	2851	**15821**
Denmark	1005	81	143	87	102	215	342	648	**2624**
Finland	920	47	328	78	189	278	238	796	**2875**
France	10096	783	3425	1161	5610	4157	3318	6852	**35401**
Germany	15712	1625	7447	2098	3314	6846	6462	12298	**55802**
Hong Kong	1499	72	176	116	304	397	766	3432	**6764**
India	562	93	354	142	546	690	1124	1063	**4574**
Indonesia	332	57	300	156	475	441	673	828	**3263**
Ireland	1081	61	235	118	124	237	149	1318	**3323**
Israel	865	49	302	247	493	265	239	706	**3166**
Italy	8404	608	3826	853	2870	5088	3080	5339	**30067**
Japan	33581	2132	5016	4104	7717	9893	11406	41285	**115133**
Malaysia	675	55	510	108	475	505	456	3626	**6411**
Netherlands	5612	382	1610	694	888	1146	1314	1991	**13638**
Norway	986	72	353	113	271	398	264	412	**2869**
Philippines	178	23	69	25	116	206	151	357	**1124**
Singapore	2959	180	703	111	349	412	736	5093	**10543**
South Africa	896	98	305	129	253	625	396	421	**3123**
South Korea	2337	205	1461	342	899	1407	2240	6562	**15453**
Spain	3299	312	844	459	574	1833	2340	1920	**11580**
Sweden	2026	141	900	214	1028	759	602	1616	**7284**
Switzerland	2506	308	1085	332	397	644	694	1042	**7008**
Taiwan	1086	97	771	237	534	774	1035	6198	**10732**
Thailand	1150	60	361	125	460	298	609	2189	**5251**
UK	11444	773	3778	1358	4585	2352	3299	7909	**35496**
USA	53509	6840	23141	9838	48114	20054	21013	57935	**240444**
TOTAL	**179954**	**16420**	**62144**	**25021**	**84653**	**65246**	**69401**	**181302**	**684142**

Notes: 1991 and 1992 are current figures at current exchange rates
1993 to 1997 are forecasts at 1992 constant values & exchange rates (ie. inflation is not included)
West Europe figures are taken from Volume 1 of the 1994 edition of the Yearbook
US, Japan & Asia Pacific figures are taken from Volume 2 of the 1994 edition of the Yearbook

SUMMARY OF MAJOR COUNTRIES MARKETS 1994

Table 2.4.4

Country	$M EDP	$M Office Equipment	$M Control & Instr	$M Medical & Industrial	$M Comms & Radar	$M Telecomm–unications	$M Consumer	$M Components	**$M TOTAL**
Australia	2820	201	770	256	855	918	952	952	**7725**
Austria	1461	126	534	209	217	487	652	974	**4661**
Belgium	3042	147	766	257	459	875	738	1392	**7677**
Brazil	5565	380	1110	430	1105	1230	2390	3752	**15961**
Canada	5111	450	1776	757	1582	1742	1985	3018	**16422**
Denmark	1036	83	147	91	113	211	352	666	**2698**
Finland	938	48	335	81	195	275	244	831	**2947**
France	10398	798	3494	1192	5722	4198	3414	7078	**36295**
Germany	16026	1657	7596	2091	3381	6709	6415	12733	**56609**
Hong Kong	1679	80	202	143	365	413	790	3481	**7153**
India	607	96	372	156	579	746	1225	1169	**4949**
Indonesia	365	58	315	175	508	450	723	953	**3549**
Ireland	1103	62	242	119	129	230	152	1402	**3439**
Israel	977	59	339	261	518	278	257	773	**3460**
Italy	8656	620	3940	885	2956	4986	3140	5591	**30773**
Japan	31566	2089	4966	4221	7794	9101	11116	41740	**112593**
Malaysia	743	58	541	120	508	515	488	4086	**7059**
Netherlands	5724	378	1594	724	905	1112	1334	2041	**13813**
Norway	1026	74	367	116	282	390	272	426	**2953**
Philippines	196	24	75	28	125	208	164	383	**1202**
Singapore	3047	184	738	124	376	408	758	5423	**11059**
South Africa	932	100	311	139	268	632	425	432	**3238**
South Korea	2430	209	1504	369	1007	1379	2307	6783	**15990**
Spain	3365	306	878	475	602	1741	2316	1939	**11623**
Sweden	1986	139	909	219	1048	736	593	1667	**7297**
Switzerland	2582	311	1096	342	409	637	713	1080	**7169**
Taiwan	1118	99	794	249	577	712	1054	6554	**11157**
Thailand	1288	65	397	135	529	322	646	2589	**5970**
UK	11787	788	3853	1416	4630	2305	3389	8364	**36532**
USA	56184	6977	23835	10206	48355	20455	21643	60500	**248155**
TOTAL	**183760**	**16665**	**63796**	**25986**	**86098**	**64404**	**70647**	**188771**	**700127**

Notes: 1991 and 1992 are current figures at current exchange rates
1993 to 1997 are forecasts at 1992 constant values & exchange rates (ie. inflation is not included)
West Europe figures are taken from Volume 1 of the 1994 edition of the Yearbook
US, Japan & Asia Pacific figures are taken from Volume 2 of the 1994 edition of the Yearbook

SUMMARY OF MAJOR COUNTRIES MARKETS 1995

Table 2.4.5

Country	$M EDP	$M Office Equipment	$M Control & Instr	$M Medical & Industrial	$M Comms & Radar	$M Telecomm–unications	$M Consumer	$M Components	**$M TOTAL**
Australia	2961	207	801	275	881	909	981	990	**8005**
Austria	1505	129	550	217	223	482	671	995	**4773**
Belgium	3164	150	796	266	472	884	757	1427	**7917**
Brazil	5899	399	1221	480	1215	1218	2533	4084	**17049**
Canada	5265	459	1847	802	1677	1760	2025	3137	**16971**
Denmark	1077	84	153	95	124	209	363	683	**2788**
Finland	967	49	345	84	200	272	251	859	**3027**
France	10814	814	3564	1223	5894	4324	3533	7277	**37444**
Germany	16827	1690	7900	2151	3482	6776	6599	13110	**58535**
Hong Kong	1813	84	222	166	420	426	811	3504	**7446**
India	668	98	391	173	613	790	1297	1269	**5299**
Indonesia	402	59	328	192	534	454	770	1066	**3806**
Ireland	1147	63	247	118	136	223	158	1449	**3541**
Israel	1094	69	376	273	538	289	273	837	**3749**
Italy	9002	632	4098	917	3045	4887	3248	5751	**31580**
Japan	32197	2152	5164	4417	8183	8919	11439	43870	**116342**
Malaysia	810	61	568	130	544	525	522	4538	**7697**
Netherlands	5953	386	1642	752	923	1123	1368	2099	**14246**
Norway	1067	75	382	120	293	382	281	437	**3037**
Philippines	220	26	82	32	137	214	181	410	**1303**
Singapore	3169	188	767	136	407	412	788	5755	**11622**
South Africa	969	101	317	150	287	638	455	448	**3364**
South Korea	2552	213	1565	399	1128	1365	2397	7060	**16679**
Spain	3466	312	913	492	626	1706	2378	1984	**11878**
Sweden	2025	141	936	226	1080	721	608	1730	**7468**
Switzerland	2711	314	1118	354	421	631	737	1116	**7401**
Taiwan	1152	101	810	261	611	669	1072	6863	**11539**
Thailand	1416	70	428	146	592	338	678	2979	**6647**
UK	12141	804	3969	1466	4723	2282	3485	8631	**37500**
USA	57308	7046	24073	10486	48838	20660	22293	62068	**252772**
TOTAL	**189761**	**16978**	**65574**	**26999**	**88249**	**64491**	**72951**	**196425**	**721429**

Notes: 1991 and 1992 are current figures at current exchange rates
1993 to 1997 are forecasts at 1992 constant values & exchange rates (ie. inflation is not included)
West Europe figures are taken from Volume 1 of the 1994 edition of the Yearbook
US, Japan & Asia Pacific figures are taken from Volume 2 of the 1994 edition of the Yearbook

SUMMARY OF MAJOR COUNTRIES MARKETS 1996

Table 2.4.6

Country	$M EDP	$M Office Equipment	$M Control & Instr	$M Medical & Industrial	$M Comms & Radar	$M Telecomm-unications	$M Consumer	$M Components	**$M TOTAL**
Australia	3080	214	833	296	907	900	1011	1023	**8264**
Austria	1550	132	561	223	228	478	684	1023	**4879**
Belgium	3259	152	820	273	482	893	773	1482	**8133**
Brazil	6253	415	1319	527	1300	1230	2650	4365	**18058**
Canada	5423	468	1903	849	1761	1777	2057	3236	**17474**
Denmark	1109	86	158	99	134	205	374	707	**2872**
Finland	996	50	355	88	206	270	257	893	**3115**
France	11247	831	3635	1256	6071	4454	3657	7557	**38707**
Germany	17837	1724	8295	2222	3621	6912	6887	13643	**61141**
Hong Kong	1922	88	235	180	454	430	836	3503	**7648**
India	721	101	410	190	644	830	1374	1377	**5648**
Indonesia	438	60	338	208	555	454	812	1167	**4032**
Ireland	1193	64	252	121	141	216	163	1510	**3661**
Israel	1204	79	414	283	554	298	288	894	**4014**
Italy	9273	645	4221	951	3136	4789	3333	5999	**32346**
Japan	34129	2216	5474	4602	8674	9098	12053	46571	**122818**
Malaysia	891	63	602	138	577	531	558	4989	**8349**
Netherlands	6192	393	1691	780	942	1134	1403	2177	**14713**
Norway	1099	77	393	124	302	379	288	454	**3115**
Philippines	241	28	89	36	148	216	197	439	**1395**
Singapore	3233	189	790	146	435	416	811	6030	**12051**
South Africa	1008	103	327	162	307	644	487	464	**3502**
South Korea	2654	218	1611	431	1218	1352	2469	7252	**17205**
Spain	3604	318	950	509	651	1689	2458	2051	**12231**
Sweden	2086	144	973	234	1123	729	636	1796	**7720**
Switzerland	2819	318	1129	365	438	637	757	1156	**7618**
Taiwan	1187	103	826	274	636	629	1090	7187	**11932**
Thailand	1530	74	454	155	651	351	710	3357	**7283**
UK	12505	812	4048	1507	4818	2304	3564	8960	**38518**
USA	59027	7187	24555	10878	49326	20866	22961	64047	**258848**
TOTAL	**197707**	**17351**	**67662**	**28108**	**90441**	**65112**	**75600**	**205309**	**747289**

Notes: 1991 and 1992 are current figures at current exchange rates
1993 to 1997 are forecasts at 1992 constant values & exchange rates (ie. inflation is not included)
West Europe figures are taken from Volume 1 of the 1994 edition of the Yearbook
US, Japan & Asia Pacific figures are taken from Volume 2 of the 1994 edition of the Yearbook

SUMMARY OF MAJOR COUNTRIES MARKETS 1997

Table 2.4.7

Country	$M EDP	$M Office Equipment	$M Control & Instr	$M Medical & Industrial	$M Comms & Radar	$M Telecomm–unications	$M Consumer	$M Components	**$M TOTAL**
Australia	3172	218	850	311	926	882	1027	1048	**8433**
Austria	1596	134	573	229	232	473	698	1052	**4988**
Belgium	3324	153	836	277	492	884	785	1539	**8289**
Brazil	6503	427	1398	570	1378	1242	2745	4616	**18879**
Canada	5531	473	1941	885	1831	1760	2090	3314	**17823**
Denmark	1131	87	161	103	145	201	383	727	**2938**
Finland	1016	51	362	91	211	267	261	924	**3183**
France	11472	847	3708	1268	6131	4365	3717	7787	**39296**
Germany	18728	1759	8627	2287	3730	6981	7094	14207	**63412**
Hong Kong	1999	90	240	189	476	426	857	3505	**7783**
India	779	104	431	207	676	863	1440	1481	**5982**
Indonesia	473	60	345	221	572	450	848	1244	**4212**
Ireland	1217	65	257	123	146	210	167	1575	**3758**
Israel	1300	87	438	291	560	301	300	942	**4220**
Italy	9458	651	4305	986	3199	4693	3386	6246	**32925**
Japan	36177	2283	5803	4794	9195	9280	12702	49465	**129698**
Malaysia	962	65	626	145	605	536	586	5407	**8931**
Netherlands	6315	397	1708	802	961	1134	1425	2244	**14987**
Norway	1121	78	405	128	311	375	294	467	**3179**
Philippines	266	29	95	40	159	216	212	466	**1483**
Singapore	3265	191	806	156	452	412	827	6258	**12367**
South Africa	1038	104	337	172	326	638	517	479	**3611**
South Korea	2734	220	1644	459	1292	1338	2534	7451	**17670**
Spain	3677	321	978	525	678	1672	2520	2105	**12476**
Sweden	2149	147	1002	240	1156	736	658	1853	**7941**
Switzerland	2904	314	1118	373	451	644	771	1190	**7764**
Taiwan	1210	104	834	284	655	598	1098	7458	**12240**
Thailand	1622	78	472	166	703	358	736	3672	**7807**
UK	12755	820	4089	1534	4914	2281	3610	9173	**39176**
USA	61388	7259	25046	11177	49820	21075	23421	65711	**264896**
TOTAL	**205281**	**17618**	**69434**	**29033**	**92381**	**65290**	**77708**	**213602**	**770347**

Notes: 1991 and 1992 are current figures at current exchange rates
1993 to 1997 are forecasts at 1992 constant values & exchange rates (ie. inflation is not included)
West Europe figures are taken from Volume 1 of the 1994 edition of the Yearbook
US, Japan & Asia Pacific figures are taken from Volume 2 of the 1994 edition of the Yearbook

3 EMERGING COUNTRIES DATA

3.1. ECONOMIC OVERVIEW

Remarkable economic progress was achieved in **China** during the nineteen eighties, with real GNP increasing by an annual average of around 10%, albeit from a low base. Economic growth has been sustained primarily by an increase in export activity. Over the past three years the rate of economic growth has accelerated. GNP increased by almost 8% in 1991 and by an estimated 12% in 1992.

A number of factors suggest that the rapid rate of economic growth will continue. With extensive deposits of raw materials and energy which are increasingly being exploited, and China seen as one of the last low wage manufacturing bases in East Asia, economic growth should continue at a fast rate. The renewal by the US in May 1994 of China's Most Favoured Nation (MFN) status will ensure continued growth of exports to the US. GNP growth is forecast to remain at over 10% per year at least until 1995.

Economy growth in **Egypt** increased by 1% in 1992 against 2.3% in 1991 and a further slow-down occurred in 1993 to 0.9%. The main export commodity, cotton, has declined sharply due to low prices and despite an increase in volumes exported from the oil sector revenue, growth has been limited by relatively depressed hydrocarbon prices. Tonnage of shipping through the Suez Canal fell in 1992 and only a modest recovery is expected in 1994. The tourist industry, which is an important source of currency, is being badly affected by terrorist attacks from fundamentalist factions.

GDP growth slowed in **Greece** from 1.8% in 1991 to 1.5% in 1992. Overall GDP growth is unlikely to exceed 1% in 1993. A slight recovery to 1.5% may occur in 1994 but higher growth will not be achieved until inflation is brought down to a lower, more sustainable level. Economic policy is firmly tied to achieving convergence with the rest of the EU in the medium term, but stabilisation measures have been needed in recent years to head off a series of incipient crises.

The rate of economic expansion in **Mexico** slackened in 1992, with a 2.2% rise in Gross Domestic Product, the lowest growth rate for four years. GDP growth is forecast to be less than 1% in 1993 but then to accelerate to around 3% in 1994, assisted by entry to NAFTA.

The **New Zealand** economy emerged from a long period of recession in the second quarter of 1991 and since then growth has been the strongest for nearly a decade. External demand continues to add impetus to the recovery and domestic demand is forecast to remain buoyant in 1994. A period of balanced growth is now in prospect with low inflation, competitive exports and a balanced external position forming the basis for a prolonged period of growth.

Following EC accession in 1986, **Portugal** became one of Europe's fastest growing economies with expansion averaging 4.6% a year. Growth has slowed sharply, however, to 1.1% in 1992 and -2.0% in 1993. Activity should pick up in 1994 and gradually gather speed in 1995 as domestic demand recovers and European export markets strengthen.

Puerto Rico's economic growth had been exceptional but recession, oil price increases and anti-inflation policies produced lower growth rates. The continued sluggishness of the

US recovery has delayed any improvements and there has been a spate of bankruptcies, and GDP in 1992 was only 1%. With the US economy now recovering from recession, in the short term the local economy should see steady growth following on from around 3% growth in the fiscal year to June 1993. In the medium term the question of statehood may again be raised with adverse effects on investment, as has been the case in the last three years, but there should be no impact on the economy in 1994.

Oil price movements and the volume of oil exports have a fundamental impact on the economic performance of **Saudi Arabia**. Lower levels of oil production in 1992 contained overall GDP growth to 5%. Lower oil prices and production have resulted in a 19% cut in government expenditure in 1993. A marginal growth of only 0.8% is expected with a 3% decline in 1994. Both the public and private sectors are likely to suffer. Real GDP growth in 1995 is projected as only 1% with the benefits of an expected rise in oil prices partially offset by a decline in production.

In 1993 the **Turkish** economy grew at a rate of 6.6% which is above the average long term growth of the economy. Both public sector expenditure and the public sector budget deficit as a share of GNP rose considerably. The high growth stimulated import demand and exports showed some growth. Excess liquidity in the market began to rise and reached very high levels in January 1994 and the Turkish Lira depreciated around 50% with very high interest rates and inflation. A sharp decline in the level of economic activity is expected in 1994.

Between 1990 and 1992, **Venezuela** enjoyed the highest rate of economic growth in Latin America, with real GDP growth averaging 8.2% per annum. However, in the first six months of 1993, real GDP growth fell to just 2.8%, compared with the same period a year earlier. With the political situation worsening the economy further slowed with negative growth of around 1% forecast for 1993. The first months of 1994 saw a fall in production and sales across a range of sectors. This is likely to continue throughout 1994 as domestic demand remains subdued.

3.2 ELECTRONICS MARKET OVERVIEW

CHINA

The market for electronic equipment and components in China is estimated at US$16.5 billion in 1992. This is about half the size of the UK market, which may appear small for a country with a population of over 1000 million. However, it should be noted that no allowance has been made for purchasing power parity, as has recently been published for economic indicators, and that the bulk of the population live in rural areas with very low incomes. It has been estimated that only 60 to 100 million are sufficiently affluent to afford foreign consumer goods.

The combined market for colour and black and white televisions is around 20 million units. The EDP market is estimated at US$2 billion, which is 18% of the UK market. The total market is forecast to grow at an average annual real rate of 10% between 1992 and 1997 with EDP by 15% p.a. and the consumer market by 5% p.a.

Production in 1992 is estimated to have been US$15.9 billion, a real growth of 14% over 1991. The real growth in 1993 is estimated at 11% and is forecast at 15% in 1994 to give a value of US$20.4 billion at 1992 exchange rates. The largest production sector remains consumer at 44% of the total in 1992. However, this proportion is falling with the growth of communications and computer equipment exceeding that of consumer goods. In 1994 computer equipment is forecast to account for 12% of the total and communications equipment for 14%. In 1994 China is forecast to produce 31 million television sets, about half of which will be colour, 2.1 million video recorders and 265 million radios and radio recorders.

China had a trade deficit in electronics in 1992 of US$0.5 billion compared with a surplus of US$1 billion in 1990, due to large imports of active components and instruments, which are required to improve the quality and technical level of production.

EGYPT

After allowing for inflation, the Egyptian market fell by 12% between 1990 and 1992 to only US$586 million. With economic problems caused by low oil prices and loss of tourist income a further fall occurred in 1993 and only modest growth is expected for the following years. Production is very limited and the country has a trade deficit of nearly US$0.5 billion.

GREECE

The Greek market for electronic goods is estimated as US$1.5 billion in 1992. Little real growth has occurred in the last 2 years and the market is still very low for a European country with a population of 10 million, emphasising its position as the poorest member of the European Union. Prospects are little better and are unlikely to improve significantly until its economy assumes a sounder position. Production of computer equipment has increased, with a real growth of 35% in 1992, but production of other electronics equipment, particularly telecommunications and colour televisions, have declined substantially with the result that total production growth was only 6% in 1992 after inflation. The country had a trade deficit of over US$1 billion in 1992.

MEXICO

Mexico is an important producer and exporter of electronic goods and has a significant home market estimated as US$7.5 billion in 1993. Main exports are computer equipment and parts, consumer equipment and components, mainly for the US market. The majority of electronic products for export operate through the Maquiladora programme, plants in Mexican territory with special export treatment.

The Mexican market enjoyed fast growth in 1991 and 1992 of 35% over the 2 years, but in 1993 experienced a decline of 1% as the rate of economic growth slackened.

Production growth was minimal in 1991 due to the depressed state of the US market but was a real 8.5% in 1992. The decline in the domestic market held growth to 2.6% in 1993 but a return to healthy growth of 5.8% is forecast for 1994.

The major export commodity remains consumer equipment with 6.4 million colour televisions exported in 1992.

NEW ZEALAND

The New Zealand market declined steeply in 1991 during the country's recession. However, recovery was strong in 1992 with a growth of 18.5% to reach US$1.2 billion. A prolonged period of growth averaging 5.9% p.a. is forecast. Production is very small and in spite of buoyant growth of 15-20% p.a. will only be US$255 million in 1994. Colour television production has virtually ceased, with the largest production sector being control and instrumentation and computer equipment, of which exports were US$54 million in 1992.

PORTUGAL

Following EC accession in 1986 Portugal experienced fast growth with similar buoyant growth in the electronics market. Growth has slowed, however, and the electronics market fell by a real 7.5% in 1992 and an estimated 4.3% in 1993. A recovery is forecast in 1994 with a return to substantial growth in 1995-97 to give a market of US$3.7 billion.

Portugal produces all types of electronic products with the most important being telecommunications, which accounts for 18% of total output; EDP equipment, mainly computer peripherals, accounting for 22% of total output; active components with 10%; passive components also 10% and consumer equipment 23%.

The industry experienced a real growth of 25% in both 1989 and 1990. However, the electronics industry went into recession in 1991 which continued through 1992 and 1993. Total reduction was around 12% after allowing for inflation. The components sector was the worst affected with a real decline of 21%. A return to growth is forecast for 1994 with a real increase of 4.7%.

PUERTO RICO

Since the previous edition of the Emerging Countries volume of the Yearbook was compiled in 1992 detailed statistics of trade between the US and Puerto Rico have become available. These indicate that the production of electronic equipment, particularly computer equipment and parts, is substantially higher than previously estimated. Total electronics production in 1992 is now estimated as US$3.1 billion of which 58% is computer goods. Other significant production is telecommunications (8%), control & instrumentation (7%), medical equipment, mainly pacemakers, (4%), and components 17%. Production fell by 12% in 1992 but growth of 5% p.a. is forecast in 1993 and 1994. The domestic market was US$1.5 billion in 1992 and is forecast to grow at an average of 5.5% between 1992 and 1997.

SAUDI ARABIA

After strong growth following the Gulf War, the Saudi Arabian electronics market declined in 1992 by 2.4%. With reduced expenditure in all sectors as a result of lower oil prices the market is forecast to continue to fall with growth not resuming until 1996, when it will only be US$1.7 million.

The Saudi Arabian electronics industry is small with the market being catered for by a large number of trading companies who import equipment and parts. However, production is increasing in some areas and significant exports are made of computer equipment, instrumentation and communications equipment totalling US$226 million.

TURKEY

The strong growth of the Turkish electronics market slowed to 7% in 1991 and in 1992 the market fell by 5% to US$3.4 billion. Growth resumed in 1993 but in early 1994 there was a sharp decline in economic activity and the market is forecast to decline again.

The Turkish electronics industry exhibited fast growth during the 1980s but in 1992 overall growth was minimal, mainly due to a decline of 10% in colour television production. EC imports from Turkey increased by 160% in 1990 but by only 21% in 1991 and fell by 2% in 1992. The majority of these imports were of colour televisions with consumer video products accounting for 73% of the total in 1992.

High demand from the former Comecom bloc, which was satisfied by Western European countries partly procured from Turkey, aided export growth but this demand has now subsided.

VENEZUELA

After the recession in 1989, growth in the electronics market resumed in 1990 and was particularly strong in 1992 with real growth of over 20%. However, the political instability in 1993 resulted in a fall in the market to US$1.5 billion due to lower imports, but production continued to expand particularly for computer equipment. Exports are minimal and the trade deficit is nearly US$1 billion.

EMERGING COUNTRIES – SUMMARY OF IMPORTS 1991

Table 3.3.1

Values in millions of current US dollars

Country	EDP	Office Equipment	Control & Instr	Medical & Industrial	Comms & Military	Telecom–munications	Consumer	Active Components	Passive Components	Other Components	TOTAL
China	668	36	850	283	462	635	205	1123	377	909	**5548**
Egypt	59	23	50	28	156	81	76	28	24	5	**529**
Greece	264	70	111	59	63	125	216	63	43	49	**1062**
Mexico	1172	177	541	144	300	588	723	484	314	743	**5186**
New Zealand	376	39	79	24	59	125	211	21	23	48	**1005**
Portugal	640	72	175	69	62	163	558	165	149	192	**2245**
Puerto Rico	421	24	114	22	27	45	63	219	106	57	**1098**
Saudi Arabia	254	16	112	78	186	112	580	23	67	110	**1538**
Turkey	407	21	219	71	117	104	91	423	191	223	**1867**
Venezuela	226	37	134	25	74	58	71	29	38	38	**730**
TOTAL	**4487**	**514**	**2385**	**802**	**1505**	**2036**	**2795**	**2577**	**1332**	**2374**	**20808**

EMERGING COUNTRIES – SUMMARY OF IMPORTS 1992

Table 3.3.2

Values in millions of current US dollars

Country	EDP	Office Equipment	Control & Instr	Medical & Industrial	Comms & Military	Telecom–munications	Consumer	Active Components	Passive Components	Other Components	TOTAL
China	985	59	1089	378	716	923	255	1449	739	1324	**7917**
Egypt	87	21	53	36	63	78	75	25	22	5	**465**
Greece	290	47	141	70	83	135	240	58	49	60	**1174**
Mexico	1422	212	734	155	420	576	896	688	358	839	**6300**
New Zealand	394	38	99	33	87	133	229	29	29	43	**1114**
Portugal	639	79	200	81	107	214	580	149	144	231	**2423**
Puerto Rico	275	27	134	37	34	103	78	288	147	67	**1190**
Saudi Arabia	278	15	139	75	152	126	559	18	80	110	**1552**
Turkey	433	20	203	77	81	89	73	404	171	170	**1721**
Venezuela	220	43	122	21	135	210	109	30	39	54	**983**
TOTAL	**5023**	**562**	**2914**	**963**	**1878**	**2586**	**3095**	**3138**	**1779**	**2903**	**24840**

Notes: Greece & Portugal figures are taken from Eurostat
Egypt, Turkey & Venezuela figures taken from UN statistics
Mexico figuers include trade with the US from Maquilodora plants
Puerto Rico estimated from trade with the US
China includes trade through Hong Kong
Saudi Arabia estimated

EMERGING COUNTRIES – SUMMARY OF EXPORTS 1991

Table 3.3.3

Values in millions of current US dollars

Country	EDP	Office Equipment	Control & Instr	Medical & Industrial	Comms & Military	Telecom-munications	Consumer	Active	Passive	Other	TOTAL
								Components			
China	314	191	70	38	178	355	2728	127	338	673	**5012**
Egypt	5	3	5	–	114	1	1	–	1	–	**130**
Greece	11	0	7	1	8	23	2	2	8	1	**63**
Mexico	589	104	242	271	200	50	1780	336	730	982	**5284**
New Zealand	60	1	28	3	22	12	1	5	5	2	**138**
Portugal	87	3	9	3	4	61	278	145	154	38	**782**
Puerto Rico	2257	–	198	100	103	56	56	21	35	247	**3073**
Saudi Arabia	35	–	51	3	69	2	–	2	5	25	**191**
Turkey	15	1	3	–	9	16	258	2	10	35	**349**
Venezuela	1	–	1	–	–	1	–	–	–	–	**3**
TOTAL	**3374**	**303**	**613**	**418**	**707**	**576**	**5104**	**641**	**1286**	**2003**	**15026**

EMERGING COUNTRIES – SUMMARY OF EXPORTS 1992

Table 3.3.4

Values in millions of current US dollars

Country	EDP	Office Equipment	Control & Instr	Medical & Industrial	Comms & Military	Telecom-munications	Consumer	Active	Passive	Other	TOTAL
								Components			
China	820	226	99	102	269	462	3357	236	749	1029	**7349**
Egypt	4	3	5	–	15	1	0	–	0	–	**28**
Greece	15	1	9	1	15	28	2	1	11	1	**84**
Mexico	559	149	372	298	190	43	2243	387	767	1028	**6036**
New Zealand	47	3	38	5	26	9	8	1	6	–	**143**
Portugal	63	2	11	5	6	65	378	147	147	42	**865**
Puerto Rico	1768	–	186	148	203	53	45	19	38	254	**2714**
Saudi Arabia	45	–	72	3	71	5	–	3	12	15	**226**
Turkey	13	1	9	2	5	19	197	2	16	40	**304**
Venezuela	1	–	3	–	1	1	–	–	–	–	**6**
TOTAL	**3336**	**385**	**804**	**564**	**801**	**685**	**6230**	**796**	**1746**	**2409**	**17755**

Notes: Greece & Portugal figures are taken from Eurostat
Egypt, Turkey & Venezuela figures taken from UN statistics
Mexico figuers include trade with the US from Maquilodora plants
Puerto Rico estimated from trade with the US
China includes trade through Hong Kong
Saudi Arabia estimated

SUMMARY OF EMERGING COUNTRIES PRODUCTION 1991

Table 3.4.1

Country	$M EDP	$M Office Equipment	$M Control & Instr	$M Medical & Industrial	$M Comms & Military	$M Telecomm–unications	$M Consumer	$M Components	$M TOTAL
China	1244	478	280	210	700	1000	6337	3414	**13663**
Egypt	24	3	2	1	15	15	71	17	**148**
Greece	36	29	16	7	30	121	79	20	**337**
Mexico	1230	66	340	373	260	275	1816	2218	**6578**
New Zealand	19	–	45	15	26	43	5	20	**173**
Portugal	339	17	90	24	118	291	316	354	**1548**
Puerto Rico	2100	–	215	138	147	110	70	592	**3372**
Saudi Arabia	86	4	80	51	75	43	10	24	**373**
Turkey	140	32	4	58	80	290	1218	157	**1979**
Venezuela	130	17	35	47	47	60	91	38	**465**
TOTAL	**5348**	**646**	**1106**	**923**	**1498**	**2247**	**10013**	**6854**	**28635**

SUMMARY OF EMERGING COUNTRIES PRODUCTION 1992

Table 3.4.2

Country	$M EDP	$M Office Equipment	$M Control & Instr	$M Medical & Industrial	$M Comms & Military	$M Telecomm–unications	$M Consumer	$M Components	$M **TOTAL**
China	1855	517	310	260	800	1230	7039	3943	**15954**
Egypt	30	4	2	1	18	17	66	18	**156**
Greece	52	53	19	7	37	121	79	25	**394**
Mexico	1310	102	350	413	250	300	2284	2298	**7307**
New Zealand	24	–	52	16	28	45	4	16	**184**
Portugal	378	16	104	28	119	304	398	364	**1711**
Puerto Rico	1770	–	200	176	246	80	64	515	**3051**
Saudi Arabia	88	4	88	59	86	48	17	28	**418**
Turkey	145	30	10	68	100	337	1113	190	**1993**
Venezuela	160	16	40	52	52	50	100	41	**511**
TOTAL	**5813**	**742**	**1175**	**1080**	**1735**	**2531**	**11164**	**7439**	**31679**

Notes: 1991 & 1992 are current figures at current exchange rates
1993 & 1994 are forecasts at 1992 constant values & exchange rates (ie. inflation is not included)

SUMMARY OF EMERGING COUNTRIES PRODUCTION 1993

Table 3.4.3

Country	$M EDP	$M Office Equipment	$M Control & Instr	$M Medical & Industrial	$M Comms & Military	$M Telecomm-unications	$M Consumer	$M Components	$M TOTAL
China	2100	541	340	310	920	1460	7509	4483	**17663**
Egypt	33	4	2	1	19	16	61	19	**156**
Greece	63	51	22	7	40	115	81	26	**406**
Mexico	1380	112	360	431	255	303	2288	2367	**7496**
New Zealand	38	–	68	17	33	48	4	16	**223**
Portugal	359	15	101	29	116	273	410	361	**1666**
Puerto Rico	1820	–	204	192	295	84	61	545	**3201**
Saudi Arabia	86	4	91	63	91	50	18	30	**433**
Turkey	160	29	11	78	120	344	1169	206	**2117**
Venezuela	176	16	42	55	55	53	106	43	**546**
TOTAL	**6214**	**773**	**1242**	**1184**	**1944**	**2746**	**11708**	**8097**	**33908**

SUMMARY OF EMERGING COUNTRIES PRODUCTION 1994

Table 3.4.4

Country	$M EDP	$M Office Equipment	$M Control & Instr	$M Medical & Industrial	$M Comms & Military	$M Telecomm-unications	$M Consumer	$M Components	$M TOTAL
China	2500	591	400	370	1100	1750	8271	5384	**20366**
Egypt	36	5	2	1	20	17	61	20	**161**
Greece	76	52	25	8	45	115	84	28	**433**
Mexico	1460	121	380	467	270	312	2414	2507	**7931**
New Zealand	47	–	82	18	38	50	4	17	**255**
Portugal	381	16	104	31	123	279	428	377	**1739**
Puerto Rico	1900	–	208	209	354	88	59	577	**3395**
Saudi Arabia	80	4	90	67	95	51	19	32	**438**
Turkey	164	29	11	81	125	351	1205	212	**2178**
Venezuela	194	17	44	58	56	54	109	45	**577**
TOTAL	**6838**	**835**	**1347**	**1310**	**2226**	**3066**	**12653**	**9199**	**37474**

Notes: 1991 & 1992 are current figures at current exchange rates
1993 & 1994 are forecasts at 1992 constant values & exchange rates (ie. inflation is not included)

SUMMARY OF MEDICAL & INDUSTRIAL PRODUCTION

Table 3.4.5

	Medical Equipment			
Country	$M	$M	$M	$M
	1991	**1992**	**1993**	**1994**
China	60	70	80	100
Egypt	1	1	1	1
Greece	1	1	1	2
Mexico	301	345	362	391
New Zealand	9	10	11	12
Portugal	16	20	21	23
Puerto Rico	100	136	150	165
Saudi Arabia	27	32	35	37
Turkey	50	60	69	72
Venezuela	25	28	30	32
TOTAL	**590**	**703**	**760**	**834**

	Industrial Equipment			
Country	$M	$M	$M	$M
	1991	**1992**	**1993**	**1994**
China	150	190	230	270
Egypt	1	1	1	1
Greece	5	6	6	6
Mexico	72	68	69	76
New Zealand	6	6	6	6
Portugal	8	8	8	9
Puerto Rico	38	40	42	44
Saudi Arabia	24	27	28	29
Turkey	8	8	9	9
Venezuela	22	24	25	26
TOTAL	**334**	**377**	**424**	**476**

Notes: 1991 & 1992 are current figures at current exchange rates
1993 & 1994 are forecasts at 1992 constant values & exchange rates (ie. inflation is not included)

SUMMARY OF CONSUMER PRODUCTION

Table 3.4.6

	Video Equipment			
Country	$M	$M	$M	$M
	1991	**1992**	**1993**	**1994**
China	3300	3440	3581	3860
Egypt	57	51	46	45
Greece	37	37	38	39
Mexico	950	1250	1275	1350
New Zealand	2	2	2	2
Portugal	118	83	85	89
Puerto Rico	48	45	43	41
Saudi Arabia	7	12	13	13
Turkey	1115	1007	1057	1089
Venezuela	38	46	51	53
TOTAL	**5672**	**5972**	**6190**	**6581**

	Audio Equipment			
Country	$M	$M	$M	$M
	1991	**1992**	**1993**	**1994**
China	2035	2384	2563	2861
Egypt	6	7	7	7
Greece	41	42	42	44
Mexico	846	1012	990	1040
New Zealand	3	2	3	2
Portugal	186	305	316	329
Puerto Rico	4	4	4	4
Saudi Arabia	3	5	5	6
Turkey	90	95	100	104
Venezuela	34	36	37	37
TOTAL	**3249**	**3891**	**4067**	**4433**

	Personal Consumer Equipment			
Country	$M	$M	$M	$M
	1991	**1992**	**1993**	**1994**
China	1002	1215	1365	1550
Egypt	8	8	9	9
Greece	2	1	1	1
Mexico	20	22	23	24
New Zealand	–	–	–	–
Portugal	11	10	9	10
Puerto Rico	18	15	14	14
Saudi Arabia	–	–	–	–
Turkey	13	11	12	12
Venezuela	19	18	18	19
TOTAL	**1092**	**1300**	**1451**	**1639**

Notes: 1991 & 1992 are current figures at current exchange rates
1993 & 1994 are forecasts at 1992 constant values & exchange rates (ie. inflation is not included)

SUMMARY OF CONSUMER PRODUCTION – UNITS

Table 3.4.7

Country	Colour Televisions–No.x1000			
	1991	**1992**	**1993**	**1994**
China	12051	13531	14000	15000
Egypt	250	200	170	180
Greece	30	30	31	33
Mexico	4900	6400	6500	6900
New Zealand	5	5	5	5
Portugal	440	300	310	330
Puerto Rico	160	210	200	190
Saudi Arabia	25	40	44	46
Turkey	2300	2185	2350	2450
Venezuela	110	125	140	150
TOTAL	**20271**	**23026**	**23750**	**25284**

Country	Monochrome Televisions–No.x1000			
	1991	**1992**	**1993**	**1994**
China	14863	15347	16000	16000
Egypt	–	–	–	–
Greece	2	4	4	3
Mexico	–	–	–	–
New Zealand	–	–	–	–
Portugal	100	110	120	130
Puerto Rico	–	–	–	–
Saudi Arabia	–	–	–	–
Turkey	–	–	–	–
Venezuela	190	250	270	280
TOTAL	**15155**	**15711**	**16394**	**16413**

Country	Video Recorders–No.x1000			
	1991	**1992**	**1993**	**1994**
China	1477	1616	1873	2100
Egypt	45	40	34	28
Greece	110	115	120	125
Mexico	–	–	–	–
New Zealand	–	–	–	–
Portugal	–	–	–	–
Puerto Rico	–	–	–	–
Saudi Arabia	–	–	–	–
Turkey	71	47	37	25
Venezuela	–	–	–	–
TOTAL	**1703**	**1818**	**2064**	**2278**

Notes: 1991 & 1992 are current figures at current exchange rates
1993 & 1994 are forecasts at 1992 constant values & exchange rates (ie. inflation is not included)

SUMMARY OF COMPONENTS PRODUCTION

Table 3.4.8

	Active Components			
Country	$M	$M	$M	$M
	1991	**1992**	**1993**	**1994**
China	1222	1285	1340	1600
Egypt	3	3	3	3
Greece	4	6	6	7
Mexico	752	770	785	848
New Zealand	5	1	1	1
Portugal	156	163	166	176
Puerto Rico	120	82	90	99
Saudi Arabia	–	–	–	–
Turkey	14	20	23	24
Venezuela	2	4	5	6
TOTAL	**2277**	**2333**	**2419**	**2764**

	Passive Components			
Country	$M	$M	$M	$M
	1991	**1992**	**1993**	**1994**
China	1130	1375	1640	1980
Egypt	4	4	5	5
Greece	12	16	16	17
Mexico	900	957	995	1045
New Zealand	10	11	12	13
Portugal	168	174	169	174
Puerto Rico	172	142	149	157
Saudi Arabia	13	16	18	19
Turkey	58	76	84	86
Venezuela	28	31	32	33
TOTAL	**2496**	**2802**	**3120**	**3529**

	Other Components			
Country	$M	$M	$M	$M
	1991	**1992**	**1993**	**1994**
China	1062	1283	1503	1804
Egypt	10	11	12	12
Greece	4	4	4	4
Mexico	566	571	587	614
New Zealand	4	4	3	3
Portugal	30	27	26	27
Puerto Rico	300	291	306	321
Saudi Arabia	11	12	13	13
Turkey	85	94	99	102
Venezuela	8	6	6	6
TOTAL	**2080**	**2303**	**2558**	**2906**

Notes: 1991 & 1992 are current figures at current exchange rates
1993 & 1994 are forecasts at 1992 constant values & exchange rates (ie. inflation is not included)

SUMMARY OF ACTIVE COMPONENTS PRODUCTION

Table 3.4.9

Product	$M 1991	$M 1992	$M 1993	$M 1994
Television Tubes & Other CRTs	254	314	365	455
Other Valves & Tubes	208	212	217	227
Diodes	91	111	114	118
Transistors	89	84	80	79
Other Discrete Semiconductors	356	345	348	391
Integrated Circuits	1279	1268	1295	1494
TOTAL	**2277**	**2333**	**2419**	**2764**

Notes: ICs includes all actives for Egypt; other discretes includes all discretes for Puerto Rico

SUMMARY OF PASSIVE COMPONENTS PRODUCTION

Table 3.4.10

Product	$M 1991	$M 1992	$M 1993	$M 1994
Capacitors	507	490	506	537
Resistors	201	197	202	208
Connectors	491	566	676	783
Chokes, Coils & Other Inductors	360	499	576	672
Relays	170	190	208	228
Switches	233	278	306	347
Printed Circuit Boards	534	582	645	754
TOTAL	**2496**	**2802**	**3120**	**3529**

Notes: PCBs included all passives for Egypt & Saudi Arabia

SUMMARY OF OTHER COMPONENTS PRODUCTION

Table 3.4.11

Product	$M 1991	$M 1992	$M 1993	$M 1994
Microphones	33	41	46	52
Loudspeakers	357	376	408	452
Amplifiers	126	127	130	135
Aerials	149	178	209	251
Unrecorded Media	668	788	921	1094
Access. & Parts for Consumer Equip.	747	794	844	923
TOTAL	**2080**	**2303**	**2558**	**2906**

Notes: accessories & parts includes all Other Components for Egypt and for saudi Arabia except for media
loadspeakers includeds microphones & amplifiers for Puerto Rico

Notes: 1991 & 1992 are current figures at current exchange rates
1993 & 1994 are forecasts at 1992 constant values & exchange rates (ie. inflation is not included)

Emerging Countries Trade 1992

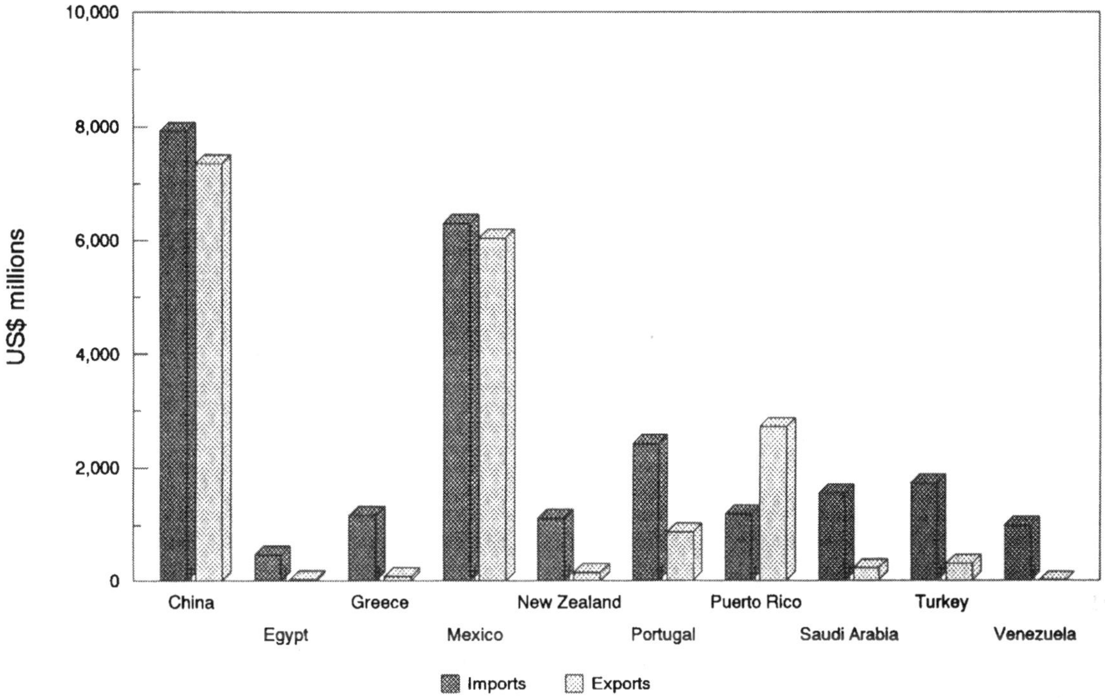

SUMMARY OF EMERGING COUNTRIES MARKETS 1991

Table 3.5.1

Country	$M EDP	$M Office Equipment	$M Control & Instr	$M Medical & Industrial	$M Comms & Military	$M Telecomm-unications	$M Consumer	$M Components	$M TOTAL
China	1598	323	1060	455	984	1280	3814	4685	**14199**
Egypt	78	23	47	29	56	95	145	66	**541**
Greece	289	98	119	65	85	222	294	164	**1337**
Mexico	1813	139	639	246	360	813	759	1711	**6480**
New Zealand	335	39	95	36	63	157	215	100	**1040**
Portugal	892	86	256	90	176	393	596	522	**3010**
Puerto Rico	264	24	131	60	71	99	77	671	**1397**
Saudi Arabia	305	20	142	126	192	153	591	191	**1720**
Turkey	532	52	220	129	188	378	1051	947	**3497**
Venezuela	355	54	168	72	121	117	162	143	**1192**
TOTAL	**6460**	**858**	**2878**	**1308**	**2296**	**3707**	**7705**	**9201**	**34412**

SUMMARY OF EMERGING COUNTRIES MARKETS 1992

Table 3.5.2

Country	$M EDP	$M Office Equipment	$M Control & Instr	$M Medical & Industrial	$M Comms & Military	$M Telecomm-unications	$M Consumer	$M Components	$M TOTAL
China	2020	350	1300	536	1247	1691	3937	5441	**16522**
Egypt	113	23	50	37	66	93	141	63	**586**
Greece	327	99	152	77	105	228	317	178	**1484**
Mexico	2173	165	712	270	480	833	937	2001	**7571**
New Zealand	371	35	113	43	90	168	225	110	**1155**
Portugal	954	93	292	104	219	452	600	554	**3268**
Puerto Rico	277	27	148	65	77	130	97	706	**1527**
Saudi Arabia	321	20	155	131	167	169	576	206	**1744**
Turkey	565	49	204	143	176	407	989	877	**3410**
Venezuela	379	59	159	73	186	259	209	164	**1488**
TOTAL	**7500**	**919**	**3286**	**1479**	**2813**	**4431**	**8028**	**10300**	**38756**

Notes: 1991 and 1992 are current figures at current exchange rates
1993 to 1997 are forecasts at 1992 constant values & exchange rates (ie. inflation is not included)

SUMMARY OF EMERGING COUNTRIES MARKETS 1993

Table 3.5.3

Country	$M EDP	$M Office Equipment	$M Control & Instr	$M Medical & Industrial	$M Comms & Military	$M Telecomm–unications	$M Consumer	$M Components	$M TOTAL
China	2549	369	1530	637	1569	2106	4160	5983	**18903**
Egypt	125	21	49	39	69	84	135	60	**582**
Greece	344	96	155	83	115	230	319	183	**1525**
Mexico	2150	168	690	278	490	808	904	2003	**7491**
New Zealand	413	39	121	54	97	150	233	128	**1236**
Portugal	896	87	281	99	215	434	587	531	**3129**
Puerto Rico	290	28	160	71	85	134	106	759	**1633**
Saudi Arabia	318	18	160	130	172	173	541	198	**1711**
Turkey	622	50	214	157	190	415	1037	916	**3601**
Venezuela	370	60	151	75	182	233	220	168	**1459**
TOTAL	**8077**	**937**	**3511**	**1622**	**3185**	**4767**	**8244**	**10929**	**41270**

SUMMARY OF EMERGING COUNTRIES MARKETS 1994

Table 3.5.4

Country	$M EDP	$M Office Equipment	$M Control & Instr	$M Medical & Industrial	$M Comms & Military	$M Telecomm–unications	$M Consumer	$M Components	**$M TOTAL**
China	2930	388	1760	740	1800	2400	4395	6657	**21070**
Egypt	134	21	48	41	74	82	138	59	**596**
Greece	371	94	161	90	126	235	327	191	**1596**
Mexico	2240	173	725	305	514	816	932	2107	**7812**
New Zealand	446	41	128	63	104	147	244	144	**1317**
Portugal	932	88	286	102	221	430	603	551	**3211**
Puerto Rico	305	28	171	76	93	137	115	817	**1742**
Saudi Arabia	307	18	156	128	175	174	510	184	**1653**
Turkey	609	49	210	163	194	399	987	934	**3545**
Venezuela	375	61	148	77	179	210	230	172	**1452**
TOTAL	**8649**	**961**	**3794**	**1784**	**3480**	**5030**	**8481**	**11815**	**43994**

Notes: 1991 and 1992 are current figures at current exchange rates
1993 to 1997 are forecasts at 1992 constant values & exchange rates (ie. inflation is not included)

SUMMARY OF EMERGING COUNTRIES MARKETS 1995

Table 3.5.5

Country	$M EDP	$M Office Equipment	$M Control & Instr	$M Medical & Industrial	$M Comms & Military	$M Telecomm–unications	$M Consumer	$M Components	**$M TOTAL**
China	3300	403	1970	860	2020	2700	4623	7306	**23182**
Egypt	143	21	49	43	77	83	142	60	**619**
Greece	408	92	171	98	138	240	343	201	**1691**
Mexico	2370	179	783	336	550	832	977	2247	**8274**
New Zealand	473	44	134	71	110	148	253	156	**1389**
Portugal	1007	91	304	110	234	439	630	579	**3394**
Puerto Rico	320	29	181	82	102	135	121	870	**1840**
Saudi Arabia	310	18	158	130	179	173	499	183	**1648**
Turkey	633	50	216	172	206	403	1007	957	**3644**
Venezuela	405	63	154	81	184	199	243	180	**1509**
TOTAL	**9369**	**990**	**4120**	**1982**	**3800**	**5352**	**8838**	**12738**	**47190**

SUMMARY OF EMERGING COUNTRIES MARKETS 1996

Table 3.5.6

Country	$M EDP	$M Office Equipment	$M Control & Instr	$M Medical & Industrial	$M Comms & Military	$M Telecomm–unications	$M Consumer	$M Components	**$M TOTAL**
China	3650	415	2170	980	2220	2900	4837	7866	**25038**
Egypt	151	22	50	45	81	84	148	62	**644**
Greece	449	91	181	106	151	244	358	211	**1792**
Mexico	2510	182	840	369	590	850	1010	2372	**8723**
New Zealand	501	45	141	81	117	150	262	167	**1465**
Portugal	1111	96	327	120	253	452	666	609	**3634**
Puerto Rico	330	29	190	88	110	133	126	919	**1925**
Saudi Arabia	324	18	164	135	186	169	510	189	**1697**
Turkey	678	51	229	184	218	411	1046	1000	**3817**
Venezuela	445	65	163	89	193	203	259	192	**1609**
TOTAL	**10149**	**1014**	**4457**	**2197**	**4120**	**5597**	**9223**	**13587**	**50343**

Notes: 1991 and 1992 are current figures at current exchange rates
1993 to 1997 are forecasts at 1992 constant values & exchange rates (ie. inflation is not included)

SUMMARY OF EMERGING COUNTRIES MARKETS 1997

Table 3.5.7

Country	$M EDP	$M Office Equipment	$M Control & Instr	$M Medical & Industrial	$M Comms & Military	$M Telecomm–unications	$M Consumer	$M Components	**$M TOTAL**
China	4000	427	2400	1110	2440	3100	5033	8447	**26957**
Egypt	160	22	52	47	85	85	154	64	**670**
Greece	493	89	192	116	166	249	374	221	**1899**
Mexico	2660	186	880	399	625	866	1046	2479	**9141**
New Zealand	531	47	148	89	124	152	272	177	**1540**
Portugal	1222	101	356	131	274	465	703	641	**3893**
Puerto Rico	340	29	198	94	117	129	131	967	**2005**
Saudi Arabia	342	19	172	141	195	171	539	199	**1779**
Turkey	725	52	243	196	233	419	1089	1055	**4012**
Venezuela	490	67	173	96	203	207	277	204	**1717**
TOTAL	**10964**	**1039**	**4814**	**2419**	**4462**	**5843**	**9617**	**14455**	**53612**

Notes: 1991 and 1992 are current figures at current exchange rates
1993 to 1997 are forecasts at 1992 constant values & exchange rates (ie. inflation is not included)

SUMMARY OF MEDICAL & INDUSTRIAL MARKETS

Table 3.5.8

	Medical Equipment						
Country	$M	$M	$M	$M	$M	$M	$M
	1991	**1992**	**1993**	**1994**	**1995**	**1996**	**1997**
China	273	321	377	450	540	640	750
Egypt	15	17	18	20	21	23	25
Greece	44	50	55	60	67	73	80
Mexico	186	197	201	221	243	267	289
New Zealand	26	31	39	46	53	61	68
Portugal	49	52	49	51	56	63	71
Puerto Rico	42	46	51	55	60	65	70
Saudi Arabia	86	86	83	80	82	86	90
Turkey	96	107	118	122	129	138	147
Venezuela	38	42	44	46	50	55	61
TOTAL	**854**	**949**	**1035**	**1151**	**1301**	**1471**	**1650**

	Industrial Equipment						
Country	$M	$M	$M	$M	$M	$M	$M
	1991	**1992**	**1993**	**1994**	**1995**	**1996**	**1997**
China	182	215	260	290	320	340	360
Egypt	14	20	20	21	22	22	23
Greece	21	27	28	30	32	33	35
Mexico	60	73	77	84	93	102	110
New Zealand	10	12	15	17	18	19	21
Portugal	41	52	49	50	54	57	60
Puerto Rico	18	19	20	21	22	23	24
Saudi Arabia	40	45	47	47	48	50	52
Turkey	33	36	39	41	43	46	49
Venezuela	34	31	31	31	31	34	35
TOTAL	**453**	**530**	**587**	**632**	**682**	**727**	**769**

Notes: 1991 and 1992 are current figures at current exchange rates
1993 to 1997 are forecasts at 1992 constant values & exchange rates (ie. inflation is not included)

SUMMARY OF CONSUMER MARKETS

Table 3.5.9

Video Equipment

Country	$M 1991	$M 1992	$M 1993	$M 1994	$M 1995	$M 1996	$M 1997
China	2716	2721	2823	2945	3078	3215	3335
Egypt	99	98	93	94	96	100	104
Greece	165	174	176	181	191	200	209
Mexico	437	566	538	554	584	605	629
New Zealand	132	134	139	146	152	158	165
Portugal	378	365	357	368	386	409	434
Puerto Rico	45	62	68	74	78	82	85
Saudi Arabia	245	241	227	214	209	215	228
Turkey	914	843	885	841	858	892	928
Venezuela	76	108	113	119	126	136	147
TOTAL	**5206**	**5312**	**5419**	**5536**	**5758**	**6013**	**6264**

Audio Equipment

Country	$M 1991	$M 1992	$M 1993	$M 1994	$M 1995	$M 1996	$M 1997
China	706	790	877	953	1015	1067	1118
Egypt	19	22	23	23	24	25	26
Greece	79	87	88	90	93	97	101
Mexico	264	307	301	310	322	332	342
New Zealand	62	72	73	75	78	80	82
Portugal	138	148	145	148	154	161	169
Puerto Rico	24	23	25	27	28	29	30
Saudi Arabia	156	153	144	135	132	135	142
Turkey	98	105	110	105	107	111	116
Venezuela	60	71	75	78	82	87	92
TOTAL	**1605**	**1778**	**1860**	**1944**	**2035**	**2125**	**2218**

Personal Consumer Eqipment

Country	$M 1991	$M 1992	$M 1993	$M 1994	$M 1995	$M 1996	$M 1997
China	392	426	460	497	530	555	580
Egypt	27	21	20	21	22	23	23
Greece	51	56	55	57	59	61	64
Mexico	58	64	65	68	71	73	75
New Zealand	21	18	22	23	23	24	25
Portugal	81	88	85	87	90	95	100
Puerto Rico	8	12	13	14	15	15	16
Saudi Arabia	190	182	171	161	158	161	169
Turkey	39	41	42	41	42	43	45
Venezuela	26	30	32	33	35	36	38
TOTAL	**894**	**939**	**965**	**1001**	**1045**	**1086**	**1135**

Notes: 1991 and 1992 are current figures at current exchange rates
1993 to 1997 are forecasts at 1992 constant values & exchange rates (ie. inflation is not included)

SUMMARY OF CONSUMER MARKETS – UNITS

Table 3.5.10

Country	Colour Televisions – No.x 1000						
	1991	**1992**	**1993**	**1994**	**1995**	**1996**	**1997**
China	9521	9404	9850	10300	10750	11150	11600
Egypt	460	438	420	425	430	440	450
Greece	367	423	425	440	470	500	520
Mexico	997	1271	1250	1300	1350	1400	1450
New Zealand	275	297	323	338	353	367	380
Portugal	849	731	725	740	770	820	870
Puerto Rico	142	150	170	190	200	210	220
Saudi Arabia	495	490	470	450	440	450	475
Turkey	1350	1410	1490	1460	1500	1580	1650
Venezuela	200	255	270	290	310	330	350
TOTAL	**14656**	**14869**	**15393**	**15933**	**16573**	**17247**	**17965**

Country	Monochrome Television – No.x 1000						
	1991	**1992**	**1993**	**1994**	**1995**	**1996**	**1997**
China	11154	10313	9110	8200	7500	7000	6600
Egypt	25	20	19	18	17	16	15
Greece	93	75	70	65	60	55	50
Mexico	230	180	170	160	155	150	150
New Zealand	20	6	6	6	6	5	5
Portugal	92	81	75	71	67	64	62
Puerto Rico	40	48	47	46	45	44	43
Saudi Arabia	included with colour televisions						
Turkey	25	20	20	19	19	19	20
Venezuela	250	330	320	310	300	290	280
TOTAL	**11929**	**11073**	**9837**	**8895**	**8169**	**7643**	**7225**

Country	Video Recorders – No.x 1000						
	1991	**1992**	**1993**	**1994**	**1995**	**1996**	**1997**
China	1205	1329	1515	1640	1770	1860	1950
Egypt	70	88	83	85	88	92	95
Greece	146	161	160	165	170	180	190
Mexico	440	520	490	500	530	550	570
New Zealand	108	112	124	130	135	141	147
Portugal	290	304	290	300	310	330	350
Puerto Rico	30	41	48	51	54	57	61
Saudi Arabia	320	300	270	250	245	250	265
Turkey	90	62	67	65	68	71	74
Venezuela	24	40	44	48	52	58	64
TOTAL	**2723**	**2957**	**3091**	**3234**	**3422**	**3589**	**3766**

Notes: 1991 and 1992 are current figures at current exchange rates
1993 to 1997 are forecasts at 1992 constant values & exchange rates (ie. inflation is not included)

SUMMARY OF COMPONENTS MARKETS

Table 3.5.11

Country	$M	$M	$M	$M	$M	$M	$M
Active Components							
	1991	**1992**	**1993**	**1994**	**1995**	**1996**	**1997**
China	2218	2498	2850	3275	3705	4085	4485
Egypt	31	28	27	26	27	28	29
Greece	64	62	64	68	73	79	85
Mexico	900	1071	1092	1158	1251	1326	1393
New Zealand	21	29	40	48	55	61	68
Portugal	175	165	162	171	185	200	216
Puerto Rico	318	351	386	425	463	500	535
Saudi Arabia	21	15	16	16	18	21	26
Turkey	435	422	443	456	470	498	533
Venezuela	31	34	36	38	42	46	51
TOTAL	**4214**	**4675**	**5114**	**5681**	**6288**	**6844**	**7420**

Country	$M	$M	$M	$M	$M	$M	$M
Passive Components							
	1991	**1992**	**1993**	**1994**	**1995**	**1996**	**1997**
China	1169	1365	1498	1648	1780	1888	1998
Egypt	27	26	25	24	25	26	26
Greece	47	54	55	57	59	62	64
Mexico	484	548	537	559	586	616	640
New Zealand	29	34	38	42	45	48	51
Portugal	163	172	165	171	180	189	199
Puerto Rico	243	251	264	277	288	296	305
Saudi Arabia	75	84	81	74	73	74	76
Turkey	239	231	240	243	247	255	265
Venezuela	66	70	73	76	79	84	89
TOTAL	**2542**	**2835**	**2976**	**3172**	**3363**	**3538**	**3713**

Country	$M	$M	$M	$M	$M	$M	$M
Other Components							
	1991	**1992**	**1993**	**1994**	**1995**	**1996**	**1997**
China	1298	1578	1635	1734	1821	1893	1964
Egypt	8	9	8	8	8	8	8
Greece	53	63	64	66	68	70	72
Mexico	327	382	374	390	410	430	446
New Zealand	50	47	51	54	56	58	59
Portugal	184	217	204	208	214	221	227
Puerto Rico	110	104	109	115	119	123	127
Saudi Arabia	96	107	102	94	92	94	97
Turkey	273	224	233	235	240	247	257
Venezuela	46	60	59	58	59	62	64
TOTAL	**2445**	**2790**	**2839**	**2962**	**3087**	**3206**	**3321**

Notes: 1991 and 1992 are current figures at current exchange rates
1993 to 1997 are forecasts at 1992 constant values & exchange rates (ie. inflation is not included)

SUMMARY OF ACTIVE COMPONENTS MARKETS

Table 3.5.12

Product	$M 1991	$M 1992	$M 1993	$M 1994	$M 1995	$M 1996	$M 1997
Television Tubes & Other CRTs	1148	1234	1303	1399	1481	1567	1654
Other Valves & Tubes	266	272	275	281	283	284	285
Diodes	87	94	97	100	105	110	114
Transistors	76	83	83	85	88	90	93
Other Discrete Semiconductors	405	456	500	543	590	637	683
Integrated Circuits	2231	2535	2858	3274	3742	4155	4591
TOTAL	**4214**	**4675**	**5114**	**5681**	**6288**	**6844**	**7420**

Notes: ICs includes all actives for Egypt & Saudi Arabias; other discretes includes all discretes for China & Puerto Rico.

SUMMARY OF PASSIVE COMPONENTS MARKETS

Table 3.5.13

Product	$M 1991	$M 1992	$M 1993	$M 1994	$M 1995	$M 1996	$M 1997
Capacitors	399	445	471	504	535	563	594
Resistors	210	229	236	247	262	269	277
Connectors	604	671	704	749	794	837	881
Chokes, Coils & Other Inductors	223	231	243	256	266	276	286
Relays	154	171	175	181	188	195	202
Switches	157	174	180	189	196	204	211
Printed Circuit Boards	796	915	966	1046	1123	1195	1262
TOTAL	**2542**	**2835**	**2976**	**3172**	**3363**	**3538**	**3713**

Notes: PCBs included all passives for Egypt & Saudi Arabia.

SUMMARY OF OTHER COMPONENTS MARKETS

Table 3.5.14

Product	$M 1991	$M 1992	$M 1993	$M 1994	$M 1995	$M 1996	$M 1997
Microphones	32	36	38	40	43	45	48
Loudspeakers	335	384	402	418	437	453	467
Amplifiers	134	145	149	157	164	172	178
Aerials	187	219	235	249	266	282	297
Unrecorded Media	453	516	516	534	552	573	592
Access. & Parts for Consumer Equip.	1305	1489	1499	1564	1626	1680	1740
TOTAL	**2445**	**2790**	**2839**	**2962**	**3087**	**3206**	**3321**

Notes: accessories & parts includes all Other Components for Egypt
loadspeakers includeds microphones & amplifiers for Puerto Rico & Saudi Arabia

Notes: 1991 and 1992 are current figures at current exchange rates
1993 to 1997 are forecasts at 1992 constant values & exchange rates (ie. inflation is not included)

4 COUNTRY DATA

4.1 CHINA

BASIC STATISTICS

Land Area:		9.6 million sq km
Population:		1140 million (1991)
Civilian Employment:		567 million (1990)
of which:	— Agriculture	60%
	— Industry	21%
	— Other	19%
GNP per Capita:		US$370 (1991) (or US$1300 based on purchasing power parity)
TV Sets in Use:		26 million
Radio Sets in Use:		20 million
Cars per 1000 Inhabitants:		2 (1992)
Main Telephone Lines per 1000 Inhabitants:		7 (1992)

LEADING ECONOMIC INDICATORS (% increases)

	Actual				Forecast		
	1988	**1989**	**1990**	**1991**	**1992**	**1993**	**1994**
Growth of GNP (Constant market prices)	11.3	3.9	5.2	7.7	12.0	13.0	11.0
Consumer Price Index	20.7	17.3	2.1	3.4	6.0	10.0	14.0
Industrial Production	20.0	6.7	6.0	14.6	19.6	13.0	na

Overview

The People's Republic of China lies in central and eastern Asia. The economy is still mainly agricultural, although production of goods, mainly consumer, has expanded in recent years. The currency of the PRC is the renminbi for internal transactions and the yuan for external, with the exchange rate now standardised.

The Peoples Republic is controlled with Chinese style socialism. The mass media and expression of critical opinion are strictly monitored. The people only pay token amounts for housing, taxes and medical care.

On an exchange rate basis per capita income is only US$370 a year but the IMF has estimated it to be around US$1300 based on purchasing power parity. This would make China the third largest economy of the world (Yearbook estimates are based on exchange rate conversions). Although the county has a population of 1140 million most are in difficult to reach rural areas and are very poor. Only 60-100 million are sufficiently affluent to afford foreign consumer goods.

GNP Growth

Remarkable economic progress was achieved during the 1980s, with real GNP increasing by an annual average of around 10%, albeit from a low base. Economic growth has been sustained primarily by an increase in export activity, and exports in the 1980s rose by an annual average of 13%. Over the past three years the rate of economic growth has accelerated. GNP increased by almost 8% in 1991 and by an estimated 12% in 1992.

A number of factors suggest that the rapid rate of economic growth will continue. There are extensive deposits of raw materials and energy which are increasingly being exploited, while China is seen as one of the last low wage manufacturing bases in East Asia. The renewal by the US in May 1994 of China's Most Favoured Nation (MFN)status will ensure continued growth of exports to the US. The overall savings ratio averages 35% while the high level of capital formation offers scope for the rapid expansion of productive capacity and productivity gains.

GNP growth is forecast to remain at over 10% per year at least until 1995.

Government Policy
The Party Congress of the Communist Party of China has confirmed the policy of reform, although the Central Committee and Politburo made it clear that they would not approve any political reorganisation. The reformist, open door economic policy is still being followed. There have been notable economic successes over the past decade and these have helped considerably to improve the availability of goods to the population. There is, however, an element of tension inherent in economic reform in the context of established political patterns, and over the long term it will not be possible to isolate the domestic, socialist economy from export oriented industry. If the pace of economic progress is not to be jeopardised, reform of the state controlled sector and system of central management will be inevitable. There are tentative signs of reform, although temporary setbacks cannot be excluded.

Prices & Inflation
Despite a rapidly growing economy, inflation in 1992 remained under control. By October the retail price index had risen by over 5% against 3.5% in the previous year but in the big cities price increases were considerably higher. As part of the policy of price reform the prices of over 500 raw materials and intermediate products were freed leading to a sharp rise in producer prices, which is estimated at around 10% for 1992. Since these price increases were passed on to consumers and the government is pressing ahead with price reform, inflation in 1993 reached double figures. Monetary expansion also continues to fuel inflation.

Overall inflation in 1994 is forecast at 14% although prices in the largest cities are climbing at more than 20% annually.

Industrial Outlook
Industrial output rose by some 20% in 1992 with, in sharp contrast, agricultural output increasing by just 3%. Around 45% of output is generated by collectively and privately owned operations and from companies established with foreign shareholders. The recent economic success has been based on a rapid expansion in exports by labour intensive, manufacturing enterprises, and on foreign direct investment, which has risen sharply in recent years.

The coastal regions predominate in economic development. The Yangtze Delta Zone with its economic centre of Shanghai and the Southern Coastal Zone with the prosperous province of Guangdong are China's economic heart. Over the last decade economic growth in Guangdong was a third higher than in China as a whole, averaging some 13% per annum, and exports account for 50% of the province's gross aggregate output. Scheduled infrastructure projects designed to link Hong Kong and Macaco more closely to mainland China this decade are expected to integrate the Guangdong province, in particular in the Pearl River Delta, with former European Crown Colonies to form a large scale economic area. This should become a major export base for China in the 21st century.

Domestic economic development has not kept pace with the growth of export orientated companies and combined with the restriction of economic momentum to individual regions this is creating imbalances, with export industry gaining at the expense of state run enterprises as a result. In the medium term, without public grants and subsidies large state owned enterprises would not be viable and the government intends to push ahead with reorganisation.

Labour Market
The country's commitment to lifetime employment in state run organisations is ending, but many workers are finding employment in the thriving private sector.

Official unemployment figures are misleading since they only apply to the state sector. The huge pool of surplus labour in the countryside which does not feature in the statistics grows remorselessly. One consequence of this is that the number of migrant labourers flocking to the east coast in search of work is as many as 80 to 90 million.

Trade & Balance of Payments
The foreign trade boom since 1991 has been a notable success. In 1992 exports grew by 18% to US$85 billion and imports by around 23% to US$78 billion, putting China amongst the worlds top ten trading nations for the first time. The balance of payments has benefitted from the export boom, with surpluses over three years totalling US$35 billion. Industrial finished products account for 75% of total exports, while imports of capital goods are gaining in importance.

Following the removal of customs barriers imports are increasing rapidly, while invisible earnings are being boosted by a further rise in tourist receipts. As a result, the trade balance has deteriorated and in 1993 and 1994 is expected to be in substantial deficit. The current account is also in deficit by as much as US$10 billion in 1993 and around US$7billion in 1994.

ELECTRONICS INDUSTRY

STRUCTURE

Number of Electronics Companies: 5060 (1992)

Main Production Activities: Consumer Electronics, Components, Computer & EDP

Major Companies:

Beijing TV Factory
China Great Wall Computer Corp
Changhong Machinery Factory
Foshan Display Equipment Factory
Hangzhou TV Factory
Nanjing Radio Factory
Qinggang Electronics Co Ltd
Shanghai Video & Audio Electronics Group Corp
Shanghai Broadcast Equipment
Shanghai Electronic Computer Factory
Shanghai Radio Factory
Shanghai TV Factory
Shanghai Radio Factory
Shoanxi Colour CRT Factory
Tianjin Telecom and Broadcasting Co.
Xinshidai Co. of China

Electronics Industry Employment: 1.67 million (1992)

In 1992, total production of the electronics industry registered a 16% real growth over 1991, to reach an estimated US$15.9 billion, compared with an 11% increase in 1991. Whilst the output of consumer products still dominates, the proportion is dropping, from 47% of the total in 1990 to a forecast 40% in 1994 as the production of computer equipment and components increases.

Production of computer equipment grew by 32% in 1991 and 49% in 1992, mainly monitors, keyboards, cases etc. Telecommunications production is also growing with high exports of handsets.

Some exporters were worried that the United States would cancel China's MFN status and were exploring other markets, Eastern Europe being particularly targeted. The removal of the threat of the loss of MFN status along with the country's re-entry to the General agreement on Tariffs and Trade (GATT) will accelerate export work.

China's electronics industry prior to the early 1980s grew slowly through a group of state-run factories under the direct jurisdiction of the relevant ministry and largely in accordance with state plans. However, production was small and work efficiency was very poor. The growth in domestic demand for consumer equipment was the corner-stone to China's acceleration in the field of electronics. The production of Chinese audio equipment, black and white televisions, and other consumer appliances soared. Since 1982, the Chinese electronics industry has been growing fairly steadily. With help from foreign experts, production improvements were achieved in business management, with the modernisation of individual factories' efficiency and production technology.

A large amount of electronics production in China still consists of assembly agreements between foreign companies and their Chinese partners. The Chinese company is supplied with the kit, helped with the purchasing of manufacturing process equipment needed to assemble the kits and the Chinese personnel are trained in manufacturing operations by the foreign partner.

Within China there are over eighty factories capable of producing mainframe, mini and microcomputers, along with a wide range of peripherals.

The two regions of Jiangsu and Shanghai account for one-third of all the production in the Chinese electronics industry. Shanghai, which includes Nanjing and several other industrial centres such as Suzhou, Wuxi, Nontong and Changzhou, has shown striking growth, further raising the degree of concentration of the production of the electronics industry in the lower Yangtze basin.

China's exporters of computer products will continue to concentrate on components, peripherals and magnetic media rather than PCs and complete systems. Foreign-investment joint ventures, particularly in the Shenzhen SEZ and other cities in the Pearl River Delta area of Guangdong province, will continue to be responsible for a large proportion of the country's computer products exports.

Like the rest of China's state-owned industries, the state-run computer sector is burdened with high costs and mounting stockpiles, and has been slow to adapt to change in the industry.

The number of makers in the industry is expected to increase as there is a continuing influx of overseas capital as well as technology from the US, Japan, Taiwan and Hong Kong. More foreign investors are expected to be attracted to Tianjin instead of the Shenzhen SEZ. By comparison, Shanghai's computer sector has been slow to develop.

The government is investing heavily in the industry by allotting funds to state run factories to upgrade their products, particularly China Great Wall. Considerable capital will be used to set up three software development and production bases in Shanghai, the Shenzhen SEZ and Beijing.

The high cost of imported components will remain the main block to exports. Most small exporters are relying on Hong Kong middlemen because of a lack of access to overseas buyers and market information.

The value of the consumer production does not reflect the large number of radios produced due to their very low cost; 158 million radios and radio combinations are produced at an average price of US$12 and 43 million tape recorders at US$7 each.

As for new products, priority will be given to developing high resolution and digital TVs, VCRs and music centres.

The importance attached to the electronics industry can be seen in the establishment of the China Electronics Corporation (CEC) in 1994. This draws together 62 subsidiaries into one company with assets of US$894. CEC will concentrate on microelectronics and telecommunication products at its plants around China. This follows the setting up of China Electronics Industry Corporation (Chinatron) in mid 1991. Chinatron is an umbrella organization for the civil and military electronics industries. Classed as a state owned national enterprise and based in Beijing, it consists of about 100 large and medium sized manufacturers, as well as many research institutes and foreign trade corporations. Its functions include market oriented R&D development of electronic products, production, marketing, contracting of projects and investment. Chinatron is headed by a president wielding ministerial powers.

Key projects for Chinatron have included the production of ICs and new lines of consumer electronics and computer products.

Among the corporation's affiliated organizations are Wuxi Microelectronics Corp., China's biggest producer of IC's; large consumer electronics conglomerates such as the Panda and Jinxling groups, and the China National Electronics Import/Export Corp. (CEIEC).

TRADE WITH THE EUROPEAN COMMUNITY

Values in current US$ millions

	Imports to the EC from China				Exports from the EC to China			
Product Group	**1989**	**1990**	**1991**	**1992**	**1989**	**1990**	**1991**	**1992**
EDP	25	40	110	250	45	34	48	39
Office Equipment	47	77	95	113	1	1	2	2
Control & Instrument	6	20	35	52	121	153	135	161
Medical	0	1	0	1	29	33	32	44
Industrial	2	9	20	32	9	17	19	6
Communications	14	25	30	33	47	89	92	62
Telecommunications	34	119	163	274	101	130	214	331
Consumer-Video	86	147	128	220	2	2	2	4
Consumer-Audio	484	770	815	870	0	1	1	1
Consumer-Personal	133	167	229	302	0	0	0	0
Active Components	2	6	10	25	7	5	10	11
Passive Components	0	0	58	94	0	0	40	28
Other Components	21	32	156	250	1	4	30	35
TOTAL	854	1413	1849	2516	363	469	625	724

Quantity in thousands of units

Product Group	**Imports to the EC from China**			**Exports from the EC to China**		
	1990	**1991**	**1992**	**1990**	**1991**	**1992**
Colour Television	891	596	787	1	1	3
Monochrome Television	97	163	190	0	0	0
Video Cassette Recorders	11	96	810	0	1	3

INDUSTRY DEVELOPMENTS

- JVC and Iwai Industrial Company of Japan have set up a joint venture with Beijing Television Equipment Plant and Beijing Comprehensive Investment Company in Beijing to produce five kinds of video cameras, initially for the Chinese market. Production is planned to reach 150 000 by 1995; localisation rate to reach 50% by 2000. The company will be known as Beijing-JVC Electronic Industrial Company Ltd.

- The Samsung Group of South Korea has established Dongguan Samsung Electronic Motor Company in Dongguan city to produce 15 million loudspeakers for audio equipment, 3 million recorder systems and 1 million computer keyboards annually. These will all be for export.

- A Russian company has set up a joint venture, Shanghai Guangsheng Company in Shanghai to design, manufacture and assemble laser and optical audio visual equipment, automatic control devices and to market the products.

- Modicon Company of the US and Shougang Corporation have established a 50:50 joint venture to produce and market Modicon computers in Beijing.

- Hitachi of Japan and other firms, Fujian Investment Enterprises Company and CNEIEC Fujian Branch have set up a joint venture called Fuijian-Hitachi TV Set Company to produce 8 mm video cameras in Fuzhou. Production is scheduled to reach 150 000 annually by 1995.

- Ricoh of Japan, Champion Technology of Hong Kong and Shanghai Facsimile Machines of China have set up a joint venture in Shanghai to make 40 000 facsimile machines a month, starting in 1996, for sale in China and to export to Asia.

- The Asian Development Bank and the Ministry of Telecommunications have made a loan of US$100 million, and a US$598 000 technical assistance grant to finance installation of two optical fibre cables and multiplexing transmission equipment for a 5500km network linking Beijing, Wuhan, Guangzhou, Taiyuan and Xi'an.

- Zhenjiang Klockner-Moeller of Germany and Zhenjiang Electric Equipment Factory of China are to supply electrical equipment worth US$450 000 for the Great Hall of the People.

- A joint venture has been established between Matsushita Electrical Industrial Company and Matsushita Electronic Components Company of Japan and Qingdao No.6 Electronics Components Plant of China to manufacture 8000 video recorders annually, starting in the first half of 1994. The new company will be known as Qingdao-Matsushita Electronic Components Company. The operation will start in the first half of 1994 and will run for 11 years.

- ABB Xiamen Switchgear Company of Switzerland and the Beijing municipality have agreed the supply of Switchgear for the Great Hall of the People.

- Nokia of Finland and the Chinese Ministry of Railways have agreed the supply of digital switching and transmission equipment worth US$7 million to Zhengzhou, Gansu and Jinan railway bureaux for use on the Baoji-Zhongwei and Jiaoxian-Huangdao lines.

- Alcatel of France and the Ministry of Posts and Telecommunications of China have formed a joint venture to be known as Alcatel of Chengdu Communication Systems to make and sell line transmission equipment.

- The World Bank is loaning the Ministry of Posts and Telecommunications of China US$250M to finance purchase of fibre optic cables, optical digital transmission equipment, telephone exchanges and power equipment, air conditioning and other equipment.

- A 50:50 joint venture has been formed by AEG of Germany and the Ministry of Metallurgical Industry Automation Research Institute of China to develop, produce and sell digital electric drive systems for the metallurgical, paper and construction materials industries. The venture will be known as Beijing AEG Electrical Drives Company.

- Zanussi of Italy has agreed a technology transfer with Chinese partners Zanussi Elettromeccanica Tianjin Compressor Company for the supply of non-CFC compressor technology.

- A joint ventue has been established in Nanjing between a Hong Kong company and Jiangsu Huashang Bar Code Co. to use Chinese technology to develop, produce and popularise laser bar code equipment.

- Matsushita Electric Industrial Company of Japan and China's 11 video recorder factories have formed a 50:50 joint venture known as Hualu-Matsushita Videocorder Company to produce 1.5 million video recorder cores a year.

- Apple Computers of the US is to set up a computer training centre in Qinghua University, Beijing, and a service centre in Beijing for distributers, developers and clients, with the co-operation of Chinese partners, the Legend Group.

- Canada's Northern Telecom is setting up two joint ventures with China's Shanghai Instrument & The Telecom Industry Bureau, one to manufacture large integrated circuits and the other a design centre for special integrated circuits for consumer electronics, telecom products and testing equipment manufacturing.

- Cony Electronics Products, part of the Conic Group of Hong Kong, has purchased a 51% stake in the Chinese company Shijiazhuang Television Factory. The facility will be developed into an export orientated electronic enterprise producing television sets and telecommunications equipment.

- AST of the US has a 90% share and Kangda Industry a 10% share of the joint venture AST China Ltd, set up in Tianjin for the production of micro computers. The company plans to produce 100 000 a year initially rising to 189 000 micros and 300 000 computer main boards after two years. Production is scheduled to begin in 1995, with 70% of production exported.

LIST OF SOURCES

Financial Times
Bank Economic Country Reports
Country Report & Country Profile — EIU
International Telecom Statistics — Siemens
World Automotive Statistics — SMMT
Predicasts F&S Index International
Electronic Times
Electronic News
Electronic Business
Electronic Engineering Times
Business China
The China Business Review
China Newsletter
China Market
China-British Trade Review
Euromonitor

Dempa Digest
Hong Kong Trade Statistics — Census of Statistics
Japan Imports & Exports — Japan Tariff Association
China Statistics Monthly
China Yearbook
Asian Sources — Computer Products
Asian Sources — Electronics
Asian Sources — Timepieces

CHINA – PRODUCTION

Table No.4.1.1

EDP		1991	1992	1993	1994	CAAGR %
Total EDP Production	YnM	6618	10240	11592	13800	
	$M	**1244**	**1855**	**2100**	**2500**	16

OFFICE EQUIPMENT		1991	1992	1993	1994	CAAGR %
Electronic Typewriters	YnM	649	668	673	679	
	$M	**122**	**121**	**122**	**123**	1
	No.th	*625*	640	680	700	
Electronic Calculators	YnM	1048	1242	1270	1380	
	$M	**197**	**225**	**230**	**250**	5
	No.th	*134000*	151000	177000	190000	
Cash Registers	YnM	48	61	77	99	
	$M	**9**	**11**	**14**	**18**	28
	No.th	*50*	*70*	*95*	*120*	
Photocopiers	YnM	798	883	966	1104	
	$M	**150**	**160**	**175**	**200**	12
	No.th	*290*	*310*	*350*	*400*	
Total Office Equipment Production	YnM	2543	2854	2986	3262	7
	$M	**478**	**517**	**541**	**591**	

CONTROL & INSTRUMENTATION		1991	1992	1993	1994	CAAGR %
Total Control & Instrumentation	YnM	1490	1711	1877	2208	
	$M	**280**	**310**	**340**	**400**	14

MEDICAL & INDUSTRIAL		1991	1992	1993	1994	CAAGR %
Total X–Ray & Medical Equipment	YnM	319	386	442	552	
	$M	**60**	**70**	**80**	**100**	20
Total Industrial Equipment	YnM	798	1049	1270	1490	
	$M	**150**	**190**	**230**	**270**	19
Total Medical & Industrial Production	YnM	1117	1435	1711	2042	
	$M	**210**	**260**	**310**	**370**	19

COMMUNICATIONS & RADAR		1991	1992	1993	1994	CAAGR %
Total Communications & Radar	YnM	3724	4416	5078	6072	
	$M	**700**	**800**	**920**	**1100**	17

TELECOMMUNICATIONS		1991	1992	1993	1994	CAAGR %
Total Telecommunications Production	YnM	5320	6790	8059	9660	
	$M	**1000**	**1230**	**1460**	**1750**	19

CHINA – PRODUCTION (cont)

CONSUMER		1991	1992	1993	1994	
Video Equipment						
Colour Television	YnM	12236	13248	13800	14904	
	$M	**2300**	**2400**	**2500**	**2700**	6
	No.th	*12051*	*13531*	*14000*	*15000*	
Monochrome Television	YnM	3405	3588	3478	3478	
	$M	**640**	**650**	**630**	**630**	−2
	No.th	*14863*	*15347*	*16000*	*16000*	
Video Recorders	YnM	1809	1987	2263	2594	
	$M	**340**	**360**	**410**	**470**	14
	No.th	*1477*	*1616*	*1873*	*2100*	
Video Cameras	YnM	106	166	226	331	
	$M	**20**	**30**	**41**	**60**	41
	No.th	*30*	*50*	*70*	*100*	
Total Video Equipment	YnM	17556	18989	19767	21307	
	$M	**3300**	**3440**	**3581**	**3860**	6
Audio Equipment						
Radios & Radio Recorders	YnM	8565	10488	11316	12696	
	$M	**1610**	**1900**	**2050**	**2300**	10
	No.th	*184000*	*218000*	*235000*	*265000*	
Tape Recorders & Players	YnM	1862	2208	2346	2539	
	$M	**350**	**400**	**425**	**460**	7
	No.th	*28737*	*32318*	*34000*	*36700*	
Record Players	YnM	266	248	226	226	
	$M	**50**	**45**	**41**	**41**	−5
	No.th	*5000*	*5000*	*4500*	*4500*	
CD Players	YnM	133	215	259	331	
	$M	**25**	**39**	**47**	**60**	24
	No.th	*360*	*600*	*720*	*900*	
Total Audio Equipment	YnM	10826	13160	14148	15793	10
	$M	**2035**	**2384**	**2563**	**2861**	
Personal Consumer Equipment						
Electronic Musical Instruments	YnM	426	552	607	662	
	$M	**80**	**100**	**110**	**120**	10
	No.th	*6000*	*7300*	*8000*	*8700*	
Electric/Electronic Watches	YnM	2724	3643	4112	4692	
	$M	**512**	**660**	**745**	**850**	13
	No.th	*446000*	*571000*	*598000*	*650000*	
Electric/Electronic Clocks	YnM	851	1049	1214	1435	
	$M	**160**	**190**	**220**	**260**	17
	No.th	*6100*	*9800*	*119000*	*140000*	
Electronic Games	YnM	1330	1463	1601	1766	
	$M	**250**	**265**	**290**	**320**	10
	No.th	*60000*	*65000*	*70000*	*77000*	
Total Personal Consumer Equipment	YnM	5331	6707	7535	8556	
	$M	**1002**	**1215**	**1365**	**1550**	13
Total Consumer Production	YnM	33713	38855	41450	45656	
	$M	**6337**	**7039**	**7509**	**8271**	8

CHINA – PRODUCTION (cont)

COMPONENTS		1991	1992	1993	1994	CAAGR %
Active Components						
Colour Television Tubes	YnM	1330	1711	1987	2484	
	$M	**250**	**310**	**360**	**450**	20
Monochrome Television Tubes	YnM	798	828	828	828	
	$M	**150**	**150**	**150**	**150**	0
Other Valves & Tubes	YnM	277	304	331	386	
	$M	**52**	**55**	**60**	**70**	13
Discrete Semiconductors	YnM	1436	1490	1490	1711	
	$M	**270**	**270**	**270**	**310**	7
ICs & Other Microcircuits	YnM	2660	2760	2760	3422	
	$M	**500**	**500**	**500**	**620**	11
Total Active Components	YnM	6501	7093	7397	8832	
	$M	**1222**	**1285**	**1340**	**1600**	12
Passive Components						
Capacitors	YnM	1064	938	966	1049	
	$M	**200**	**170**	**175**	**190**	6
Resistors	YnM	479	497	524	552	
	$M	**90**	**90**	**95**	**100**	5
Connectors	YnM	1224	1656	2208	2760	
	$M	**230**	**300**	**400**	**500**	29
Switches	YnM	452	607	773	994	
	$M	**85**	**110**	**140**	**180**	28
Relays	YnM	372	497	607	718	
	$M	**70**	**90**	**110**	**130**	20
Inductors	YnM	612	1297	1656	2098	
	$M	**115**	**235**	**300**	**380**	27
Printed Circuit Boards	YnM	1809	2098	2318	2760	
	$M	**340**	**380**	**420**	**500**	15
Total Passive Components	YnM	6012	7590	9053	10930	
	$M	**1130**	**1375**	**1640**	**1980**	20
Other Components						
Microphones	YnM	90	127	155	188	
	$M	**17**	**23**	**28**	**34**	22
Loudspeakers	YnM	1436	1601	1766	1987	
	$M	**270**	**290**	**320**	**360**	11
Amplifiers	YnM	399	442	469	497	
	$M	**75**	**80**	**85**	**90**	6
Aerials	YnM	638	828	994	1214	
	$M	**120**	**150**	**180**	**220**	21
Unrecorded Media	YnM	1490	2208	2815	3643	
	$M	**280**	**400**	**510**	**660**	28
Accessories & Parts for Consumer Equipment	YnM	1596	1877	2098	2429	
	$M	**300**	**340**	**380**	**440**	14
Total Other Components	YnM	5650	7082	8297	9958	
	$M	**1062**	**1283**	**1503**	**1804**	19
Total Components Production	YnM	18162	21765	24746	29720	
	$M	**3414**	**3943**	**4483**	**5384**	17

CHINA – PRODUCTION (cont)

		1991	1992	1993	1994	CAAGR %
TOTAL PRODUCTION–CHINA	YnM	72687	88066	97500	112420	
	$M	**13663**	**15954**	**17663**	**20366**	13

Notes: 1991 and 1992 are current figures at current exchange rates
1993 & 1994 are forecasts at 1992 constant values & exchange rates (ie inflation is not included)
Current estimates & forecasts are made in US$ with local values shown for guidance only
CAAGR column indicates true annual average growth between 1992 & 1994

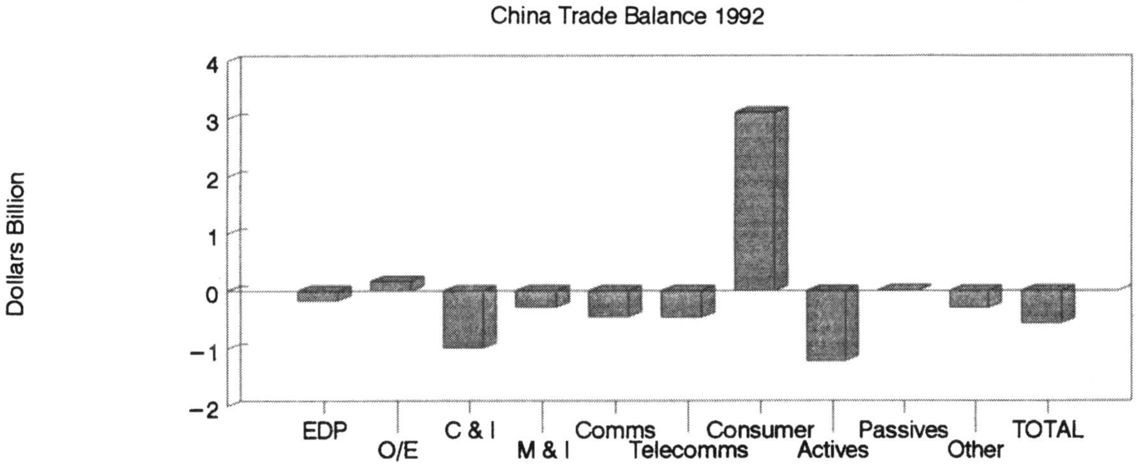

CHINA – MARKETS

Table No.4.1.2

EDP		**1991**	**1992**	**1993**	**1994**	**1995**	**1996**	**1997**	CAAGR %
Total EDP Market	YnM	8501	11150	14070	16174	18216	20148	22080	
	$M	**1598**	**2020**	**2549**	**2930**	**3300**	**3650**	**4000**	15

OFFICE EQUIPMENT		**1991**	**1992**	**1993**	**1994**	**1995**	**1996**	**1997**	CAAGR %
Electronic Typewriters	YnM	638	679	679	668	657	646	635	
	$M	**120**	**123**	**123**	**121**	**119**	**117**	**115**	-1
	No.th	*620*	*639*	*678*	*685*	*690*	*695*	*700*	
Electronic Calculators	YnM	202	221	232	243	254	259	265	
	$M	**38**	**40**	**42**	**44**	**46**	**47**	**48**	4
	No.th	*25030*	*26186*	*27160*	*28400*	*29500*	*30100*	*30700*	
Cash Registers	YnM	74	110	121	127	132	138	144	
	$M	**14**	**20**	**22**	**23**	**24**	**25**	**26**	5
	No.th	*44*	*51*	*59*	*62*	*65*	*68*	*70*	
Photocopiers	YnM	803	922	1005	1104	1181	1248	1314	
	$M	**151**	**167**	**182**	**200**	**214**	**226**	**238**	7
	No.th	*269*	*292*	*334*	*370*	*400*	*420*	*440*	
Total Office Equipment Market	YnM	1718	1932	2037	2142	2225	2291	2357	
	$M	**323**	**350**	**369**	**388**	**403**	**415**	**427**	4

CONTROL & INSTRUMENTATION		**1991**	**1992**	**1993**	**1994**	**1995**	**1996**	**1997**	CAAGR %
Total Control & Instrumentation	YnM	5639	7176	8446	9715	10874	11978	13248	
	$M	**1060**	**1300**	**1530**	**1760**	**1970**	**2170**	**2400**	13

MEDICAL & INDUSTRIAL		**1991**	**1992**	**1993**	**1994**	**1995**	**1996**	**1997**	CAAGR %
Total X–Ray & Medical Equipmen	YnM	1452	1772	2081	2484	2981	3533	4140	
	$M	**273**	**321**	**377**	**450**	**540**	**640**	**750**	18
Total Industrial Equipment	YnM	968	1187	1435	1601	1766	1877	1987	
	$M	**182**	**215**	**260**	**290**	**320**	**340**	**360**	11
Total Medical & Industrial Market	YnM	2421	2959	3516	4085	4747	5410	6127	
	$M	**455**	**536**	**637**	**740**	**860**	**980**	**1110**	16

COMMUNICATIONS & RADAR		**1991**	**1992**	**1993**	**1994**	**1995**	**1996**	**1997**	CAAGR %
Total Communications & Radar	YnM	5235	6883	8661	9936	11150	12254	13469	
	$M	**984**	**1247**	**1569**	**1800**	**2020**	**2220**	**2440**	14

TELECOMMUNICATIONS		**1991**	**1992**	**1993**	**1994**	**1995**	**1996**	**1997**	CAAGR %
Total Telecommunications Market	YnM	6810	9334	11625	13248	14904	16008	17112	
	$M	**1280**	**1691**	**2106**	**2400**	**2700**	**2900**	**3100**	13

CONSUMER		**1991**	**1992**	**1993**	**1994**	**1995**	**1996**	**1997**	CAAGR %
Video Equipment									
Colour Television	YnM	10177	10632	11255	11868	12420	12972	13524	
	$M	**1913**	**1926**	**2039**	**2150**	**2250**	**2350**	**2450**	5
	No.th	*9521*	*9404*	*9850*	*10300*	*10750*	*11150*	*11600*	

CHINA – MARKETS (cont)

		1991	1992	1993	1994	1995	1996	1997	
Monochrome Television	YnM	2649	2556	2252	2042	1904	1794	1711	
	$M	**498**	**463**	**408**	**370**	**345**	**325**	**310**	-8
	No.th	*11154*	*10313*	*9110*	*8200*	*7500*	*7000*	*6600*	
Video Recorders	YnM	1490	1650	1827	1987	2153	2263	2374	
	$M	**280**	**299**	**331**	**360**	**390**	**410**	**430**	8
	No.th	*1205*	*1329*	*1515*	*1640*	*1770*	*1860*	*1950*	
Video Cameras & Camera Combinations	YnM	133	182	248	359	513	718	800	
	$M	**25**	**33**	**45**	**65**	**93**	**130**	**145**	34
	No.th	*36*	*54*	*75*	*110*	*160*	*230*	*260*	
Total Video Equipment	YnM	14449	15020	15583	16256	16991	17747	18409	
	$M	**2716**	**2721**	**2823**	**2945**	**3078**	**3215**	**3335**	4
Audio Equipment									
Radios & Radio Recorders	YnM	2128	2512	2870	3146	3367	3533	3698	
	$M	**400**	**455**	**520**	**570**	**610**	**640**	**670**	8
	No.th	*32000*	*36964*	*42100*	*46000*	*49000*	*51000*	*53000*	
Tape Recorders & Players	YnM	1373	1568	1689	1822	1932	2042	2153	
	$M	**258**	**284**	**306**	**330**	**350**	**370**	**390**	7
	No.th	*19987*	*21303*	*23000*	*24700*	*26200*	*27700*	*29200*	
Record Players	YnM	223	232	221	215	210	204	199	
	$M	**42**	**42**	**40**	**39**	**38**	**37**	**36**	-3
	No.th	*4858*	*4794*	*4600*	*4500*	*4400*	*4300*	*4200*	
Compact Disc Players	YnM	32	50	61	77	94	110	121	
	$M	**6**	**9**	**11**	**14**	**17**	**20**	**22**	20
	No.th	*67*	*100*	*120*	*155*	*190*	*230*	*260*	
Total Audio Equipment	YnM	3756	4361	4841	5261	5603	5890	6171	
	$M	**706**	**790**	**877**	**953**	**1015**	**1067**	**1118**	7
Personal Consumer Equipment									
Electronic Musical Instruments	YnM	351	397	442	469	497	524	552	
	$M	**66**	**72**	**80**	**85**	**90**	**95**	**100**	7
	No.th	*5355*	*5745*	*6300*	*6700*	*7000*	*7300*	*7600*	
Electric/Electronic Watches	YnM	755	889	972	1076	1159	1214	1270	
	$M	**142**	**161**	**176**	**195**	**210**	**220**	**230**	7
	No.th	*83000*	*87079*	*91830*	*100000*	*107000*	*111000*	*115000*	
Electric/Electronic Clocks	YnM	484	535	574	618	662	690	718	
	$M	**91**	**97**	**104**	**112**	**120**	**125**	**130**	6
	No.th	*31319*	*33183*	*35200*	*37500*	*39700*	*41300*	*42800*	
Electronic Games	YnM	495	530	552	580	607	635	662	
	$M	**93**	**96**	**100**	**105**	**110**	**115**	**120**	5
	No.th	*21903*	*23500*	*25000*	*26500*	*28000*	*29500*	*31000*	
Total Personal Equipment	YnM	2085	2352	2539	2743	2926	3064	3202	
	$M	**392**	**426**	**460**	**497**	**530**	**555**	**580**	6
Total Consumer Market	YnM	20290	21732	22963	24260	25519	26700	27782	
	$M	**3814**	**3937**	**4160**	**4395**	**4623**	**4837**	**5033**	5

COMPONENTS		1991	1992	1993	1994	1995	1996	1997	CAAGR %
Active Components									
Colour Television Tubes	YnM	3277	3610	3870	4195	4471	4747	5023	
	$M	**616**	**654**	**701**	**760**	**810**	**860**	**910**	7
Monochrome Television Tubes	YnM	857	905	889	883	856	828	800	
	$M	**161**	**164**	**161**	**160**	**155**	**150**	**145**	-2
Other Valves & Tubes	YnM	293	348	381	414	442	469	497	
	$M	**55**	**63**	**69**	**75**	**80**	**85**	**90**	7

CHINA – MARKETS (cont)

		1991	1992	1993	1994	1995	1996	1997	
Discrete Semiconductors	YnM	1543	1833	2053	2263	2484	2705	2926	
	$M	**290**	**332**	**372**	**410**	**450**	**490**	**530**	10
ICs & Other Microcircuits	YnM	5831	7093	8539	10322	12199	13800	15511	
	$M	**1096**	**1285**	**1547**	**1870**	**2210**	**2500**	**2810**	17
Total Active Components	YnM	11800	13789	15732	18078	20452	22549	24757	
	$M	**2218**	**2498**	**2850**	**3275**	**3705**	**4085**	**4485**	12
Passive Components									
Capacitors	YnM	1016	1225	1352	1490	1601	1711	1822	
	$M	**191**	**222**	**245**	**270**	**290**	**310**	**330**	8
Resistors	YnM	585	707	745	800	856	883	911	
	$M	**110**	**128**	**135**	**145**	**155**	**160**	**165**	5
Connectors	YnM	1287	1551	1711	1877	2026	2153	2291	
	$M	**242**	**281**	**310**	**340**	**367**	**390**	**415**	8
Switches	YnM	516	607	651	690	718	745	773	
	$M	**97**	**110**	**118**	**125**	**130**	**135**	**140**	5
Relays	YnM	362	420	442	469	497	519	541	
	$M	**68**	**76**	**80**	**85**	**90**	**94**	**98**	5
Inductors	YnM	559	651	718	773	817	850	883	
	$M	**105**	**118**	**130**	**140**	**148**	**154**	**160**	6
Printed Circuit Boards	YnM	1894	2374	2650	2997	3312	3560	3809	
	$M	**356**	**430**	**480**	**543**	**600**	**645**	**690**	10
Total Passive Components	YnM	6219	7535	8269	9097	9826	10422	11029	
	$M	**1169**	**1365**	**1498**	**1648**	**1780**	**1888**	**1998**	8
Other Components									
Microphones	YnM	106	121	132	144	155	166	177	
	$M	**20**	**22**	**24**	**26**	**28**	**30**	**32**	8
Loudspeakers	YnM	1091	1270	1380	1463	1546	1601	1656	
	$M	**205**	**230**	**250**	**265**	**280**	**290**	**300**	5
Amplifiers	YnM	420	475	508	541	569	596	618	
	$M	**79**	**86**	**92**	**98**	**103**	**108**	**112**	5
Aerials	YnM	479	569	635	690	745	800	856	
	$M	**90**	**103**	**115**	**125**	**135**	**145**	**155**	9
Unrecorded Media	YnM	851	1005	1016	1049	1076	1104	1132	
	$M	**160**	**182**	**184**	**190**	**195**	**200**	**205**	2
Access & Parts for Consumer Equip	YnM	3958	5272	5354	5686	5962	6182	6403	
	$M	**744**	**955**	**970**	**1030**	**1080**	**1120**	**1160**	4
Total Other Components	YnM	6905	8711	9025	9572	10052	10449	10841	
	$M	**1298**	**1578**	**1635**	**1734**	**1821**	**1893**	**1964**	4
Total Components Market	YnM	24924	30034	33026	36747	40329	43420	46627	
	$M	**4685**	**5441**	**5983**	**6657**	**7306**	**7866**	**8447**	9

		1991	1992	1993	1994	1995	1996	1997	CAAGR
TOTAL MARKET – CHINA	YnM	75539	91201	104345	116306	127965	138210	148803	
	$M	**14199**	**16522**	**18903**	**21070**	**23182**	**25038**	**26957**	10

Notes: 1991 and 1992 are current figures at current exchange rates
1993 to 1997 are forecasts at 1992 constant values & exchange rates (ie. inflation is not included)
Current estimates & forecasts are made in US$ with local values shown for guidance only
CAAGR column indicates true annual average growth between 1992 & 1997

4.2 EGYPT

BASIC STATISTICS

Land Area:		1001 450 sq km
Population:		58 million (1992)
Civilian Employment (target)		14.3 million (1991)
of which:	- Agriculture	34%
	- Manufacturing & Mining	16%
	- Other	50%
GDP per Capita:		US$630 (1992 estimate)
TV Sets in Use:		5 million
Radio Sets in Use:		14 million
Cars per 1000 Inhabitants:		21 (1992)
Main Telephones Lines per 1000 Inhabitants:		36 (1992)

LEADING ECONOMIC INDICATORS (% increases)

	Actual				Forecast		
	1988	**1989**	**1990**	**1991**	**1992**	**1993**	**1994**
Growth of GDP							
(constant market prices)	3.9	3.0	2.6	2.3	1.0	0.9	1.5
Consumer Price Index	17.6	21.3	16.7	20.7	9.7	12.0	10.0
Industrial Production	7.0	7.3	6.5	6.0	4.9	na	na

Overview

The Arab Republic of Egypt occupies the north-eastern corner of Africa, most of the country is covered by the vast Western and Eastern Deserts. Until June 1954, Egypt was a constitutional monarchy under Britain. The country's four main sources of revenue are oil sales, Suez Canal tolls, tourism receipts and workers remittances. Egypt's economy has suffered in recent years with the decline in oil prices, and it has turned to foreign borrowing and aid to help to finance itself. Debt servicing remains a problem.

GDP Growth

Economy growth increased by 1% in 1992 against 2.3% in 1991 and a further slow-down occurred in 1993. The main export commodity, cotton, has declined sharply due to low prices and despite an increase in volumes exported from the oil sector revenue growth has been limited by relatively depressed hydrocarbon prices. Tonnage of shipping through the Suez Canal fell in 1992 and only a modest recovery is expected in 1994. The tourist industry, an important source of currency, is being badly affected by terrorist attacks from fundamentalist factions.

Government Policy

In Egypt the political system is based around the People's Assembly which is a legislative body and approves general policy, the budget and the development plan. The People's Assembly nominates the president, whose candidature is submitted to public referendum. In the April 1987 elections, Hasni Mubarak's National Democratic Party won with a large majority and in the elections held in November 1990, the NDP again won an overwhelming majority.

Government policy is mainly centred around the need to cut foreign debt. Following receipt of compensation from the allies after the Gulf War the debt was cut to US$41.5 billion. However, an accord with the World Bank was delayed because the two parties were unable to agree a schedule for further economic measures including, in particular, an acceleration of the privatisation process. The government is being careful that economic policies do not aggravate social unrest or increase radical demands.

Prices & Inflation

Like so many other countries, Egypt managed relatively modest price inflation before 1973 (about 3% annually over a ten year period). Another era was entered from 1973 to 1980 with inflation levels of 11–14%, rising to around 24% in 1986.

Subsidies in Egypt are both hidden and open, intentional and unintentional. Direct control of prices of basic commodities to protect the poor is a legacy from Nasser's time but the cost then was small. In 1991/92 expenditure of E£4.5 billion was budgeted on subsidies.

Inflation is estimated to have fallen to 9.7% in 1992 on the basis of a significant fall in investment associated with the economic downturn, and financing the government borrowing requirement through the issue of treasury bills, which contributes to a slower growth in the money supply. Inflation, however, is expected to rise to 12% in 1993.

Industrial Outlook

State owned enterprises continue to dominate trade and industry, with some two thirds of the industrial companies and the largest banks under centralised control. Although privatisation is a declared goal of the government, progress in the area has been hampered by political difficulties, although in recent years there has been an expansion in the private sector in the areas of commerce and services. Following restructuring of the state owned sector there are now twenty seven new holding companies. Ten companies have been selected to date, and a further twenty five will be selected for privatisation by mid year. It is planned to remove the state monopoly in cotton and sugar cane by 1994. The bulk of the overseas funds made available to finance the privatisation process, totalling E£1.5 billion, has not been drawn on because of the slow rate of progress.

Labour Market

The population is increasing at an annual rate of 2.5%, equivalent to some 1.3 million people. The rate of unemployment has reached about 20% of the labour force and social problems have not been eased by the return of about 600 000 guest workers from Iraq and Kuwait.

A government guarantee to provide work for all university graduates has produced a large surplus of underemployed, ill paid civil servants many of whom have to take a second job in the informal sector.

Trade & Balance of Payments

Dependence on food imports is running at US$8.5 billion a year and is to be reduced by removing many of the restrictions on the cultivation of grain. Other important imports are transport equipment and machinery, chemicals, timber and paper, while key exports are cotton, oil and oil derivatives products, industrial commodities and agricultural products.

The loss of eastern European markets has particularly affected exports of cotton and reflected in a 40% decline in revenue over 1990-1991, weakening the dominant role of cotton as an export commodity. Import bans or restrictions remain in force, but the former are gradually being replaced by import duties. Restrictions on items which are also manufactured locally will remain in force for the foreseeable future.

The trade deficit is forecast to remain at US$5 billion to US$6 billion a year but the current account showed a surplus of US$2.8 billion in 1992 due to transfers. Official transfers will keep it in surplus in 1993 and 1994.

ELECTRONICS INDUSTRY

STRUCTURE

Main Production Activities: Colour Television Sets, Monochrome Television Sets, Communications and Military Equipment, EDP.

Major Companies:

- Benha Electronics Company
- Egyptian Elecronics Industry
- Electrical Manufacturing Group/EMG
- Factory Electronics Technology/FET
- Fresh/National Engineering Industries
- Future Computers
- Goldstar Egypt Electronics
- Giza Sysrem Engineering
- Maasara Engineering Industries
- Nasr Television & Electronics Co.
- Nasr Co. for Electrical & Electronic Apparatus (Philips)
- Nasr Co. for Transformers & Computers
- Nekhily & Bros Engineering
- Ratheon/Mediterranian Systems
- TELEMISR/Arab Radio Trans & Elactronics
- XEROX EGYPT

The market for computers and data processing equipment is growing strongly having been largely fuelled on an initiative, started by presidential decree in 1981, to computerise data bases in all government and public sector organisations. The mid-1980s saw an increase in the use of microcomputers and the sharp increase in private sector usage. In 1992 a contract worth US$150 million was awarded for the automation of the port of Alexandria. In response to this growth Group Bull established a branch office in Cairo in 1992. Bull is the forth computer company to base a subsidiary in Cairo after ICL, NCR and IBM.

Locally-produced computers are only in the form of PC compatibles assembled by the Benha Electronics Company, which is part of the Ministry of Military Production, exclusively for the Armed Forces. Egyptian Advanced Technologies has been marketing its own BALSAM brand PC compatibles, made to its own specification in the Far East, for a few years now. Local production is aimed at the lower end of the market, with modest specifications, and prices are expected to under-cut products imported in assembled from from the Far East. Software is also produced locally and Bull is to establish a new factory to produce Arabic software for the Middle east.

Egypt is very strong in the sales of military electronics to the Middle East and Africa, surpassing China, Brazil, Israel and Singapore. The main Egyptian company operating in this area is the Benha Company for Electronics Industries which is situated 55 km north of Cairo, employing over 3000 people and having an annual turnover in excess of US$50 million, mainly in consumer equipment. Benha is now the biggest electronics company in the Middle East. At present the company also supplies Plessey of the UK with tank radios, squad radios and ground-to-ground radio jammers. Egypt imports large quantities of radar and radio equipment much of which is thought to be re-exported.

Under Year 2 of the five-year economic plan the electronics industry in Egypt was greatly encouraged. Industrial projects in which the private sector was encouraged were primarily components for televisions and radios, and the encouragement of production and assembly of personal computers. In 1987/88, 54 electronics industry projects were approved, comprising 8.8% of the total new projects, with investment costs of E£213 million.

Expansion and upgrading of telecommunications continues to be one of the most important infrastructure needs in Egypt. In the past, due to lack of financial resources, telecommunications had been a neglected sector. At present, local production supplies about 40% of Egypt's demands for wires and cables as well as some telephone sets.

TRADE WITH THE EUROPEAN COMMUNITY

Values in current US$ millions

	Imports to the EC from Egypt				Exports from the EC to Egypt			
Product Group	**1989**	**1990**	**1991**	**1992**	**1989**	**1990**	**1991**	**1992**
EDP	1	1	2	2	31	40	37	31
Office Equipment	0	0	0	1	5	6	5	6
Control & Instruments	0	5	4	4	37	58	42	44
Medical	0	0	0	0	6	10	13	15
Industrial	0	0	0	0	5	21	13	17
Communications	2	5	5	6	71	173	130	37
Telecommunications	0	1	1	0	34	83	57	54
Consumer-Video	0	0	1	0	1	2	2	3
Consumer-Audio	0	3	0	0	0	1	2	1
Consumer-Personal	0	0	0	0	1	1	0	1
Active Components	0	2	1	0	1	6	5	6
Passive Components	0	0	0	0	6	12	22	20
Other Components	0	0	0	15	2	4	95	43
TOTAL	3	17	14	28	200	417	423	278

Quantity in thousands of units

	Imports to the EC from Egypt		Exports from the EC to Egypt	
Product Group	**1991**	**1992**	**1991**	**1992**
Colour Television	0	0	1	2
Monochrome Television	0	0	0	0
Video Cassette Recorders	0	0	0	1

LIST OF SOURCES

Financial Times
Bank Economic Country Reports
Country Report & Country Profile — EIU
International Telecom Statistics — Siemens
World Automotive Statistics — SMMT
Predicasts F&S Index International
Electronic Times
Electronic News
Electronic Business
Electronic Engineering Times
Computer User
Euromonitor
Egypt Trade
Egypt Directory of Service — Industrial & Foreign Trade Co

EGYPT – PRODUCTION

Table No.4.2.1

EDP		1991	1992	1993	1994	CAAGR %
Total EDP Production	E£M	80	100	110	120	10
	$M	**24**	**30**	**33**	**36**	

OFFICE EQUIPMENT		1991	1992	1993	1994	CAAGR %
Electronic Calculators,	E£M	5	6	6	6	0
	$M	**2**	**2**	**2**	**2**	
	No.th	*400*	500	510	520	
Photocopiers	E£M	6	7	8	9	13
	$M	**2**	**2**	**2**	**3**	
Total Office Equipment Production	E£M	11	13	14	15	7
	$M	**3**	**4**	**4**	**5**	

CONTROL & INSTRUMENTATION		1991	1992	1993	1994	CAAGR %
Total Control & Instrumentation	E£M	6	7	7	8	7
	$M	**2**	**2**	**2**	**2**	

MEDICAL & INDUSTRIAL		1991	1992	1993	1994	CAAGR %
Total X–Ray & Medical Equipment	E£M	2	2	2	2	0
	$M	**1**	**1**	**1**	**1**	
Total Industrial Equipment	E£M	2	2	2	2	0
	$M	**1**	**1**	**1**	**1**	
Total Medical & Industrial Production	E£M	4	4	4	4	0
	$M	**1**	**1**	**1**	**1**	

COMMUNICATIONS & RADAR		1991	1992	1993	1994	CAAGR %
Total Communications & Radar	E£M	50	60	63	67	6
	$M	**15**	**18**	**19**	**20**	

TELECOMMUNICATIONS		1991	1992	1993	1994	CAAGR %
Total Telecommunications Production	E£M	50	55	54	55	0
	$M	**15**	**17**	**16**	**17**	

EGYPT – PRODUCTION (cont)

CONSUMER		1991	1992	1993	1994	
Video Equipment						
Colour Television	E£M	150	140	128	130	-4
	$M	**45**	**42**	**38**	**39**	
	No.th	*250*	*200*	*170*	*180*	
Video Recorders	E£M	40	30	25	20	-18
	$M	**12**	**9**	**8**	**6**	
	No.th	*45*	*40*	*34*	*28*	
Total Video Equipment	E£M	190	170	153	150	-6
	$M	**57**	**51**	**46**	**45**	
Audio Equipment						
Radio Sets	E£M	10	14	14	14	0
	$M	**3**	**4**	**4**	**4**	
	No.th	*100*	*150*	*145*	*140*	
Tape Recorders, Record Players & CD Players	E£M	10	8	8	9	6
	$M	**3**	**2**	**2**	**3**	
	No.th	*85*	*50*	*51*	*54*	
Total Audio Equipment	E£M	20	22	22	23	2
	$M	**6**	**7**	**7**	**7**	
Personal Consumer Equipment						
Electronic Musical Instruments	E£M	2	2	2	2	0
	$M	**1**	**1**	**1**	**1**	
	No.th	*20*	*18*	*18*	*19*	
Electric/Electronic Watches	E£M	20	22	23	23	2
	$M	**6**	**7**	**7**	**7**	
	No.th	*1000*	*1100*	*1105*	*1200*	
Electric/Electronic Clocks	E£M	3	4	4	4	0
	$M	**1**	**1**	**1**	**1**	
	No.th	*200*	*300*	*280*	*300*	
Total Personal Consumer Equipment	E£M	25	28	29	29	2
	$M	**8**	**8**	**9**	**9**	
Total Consumer Production	E£M	235	220	204	202	-4
	$M	**71**	**66**	**61**	**61**	

COMPONENTS		1991	1992	1993	1994	CAAGR %
Active Components						
Total Active Components	E£M	10	10	10	11	5
	$M	**3**	**3**	**3**	**3**	
Passive Components						
Total Passive Components	E£M	12	14	15	15	4
	$M	**4**	**4**	**5**	**5**	
Other Components						
Total Other Components	E£M	12	13	14	14	4
	$M	**4**	**4**	**4**	**4**	
Total Components Production	E£M	34	37	39	40	4
	$M	**10**	**11**	**12**	**12**	

EGYPT – PRODUCTION (cont)

		1991	1992	1993	1994	CAAGR %
TOTAL PRODUCTION–EGYPT	E£M	470	496	495	511	2
	$M	**141**	**149**	**149**	**153**	

Notes: 1991 and 1992 are current figures at current exchange rates
1993 & 1994 are forecasts at 1992 constant values & exchange rates (ie inflation is not included)
CAAGR column indicates true annual average growth between 1992 & 1994

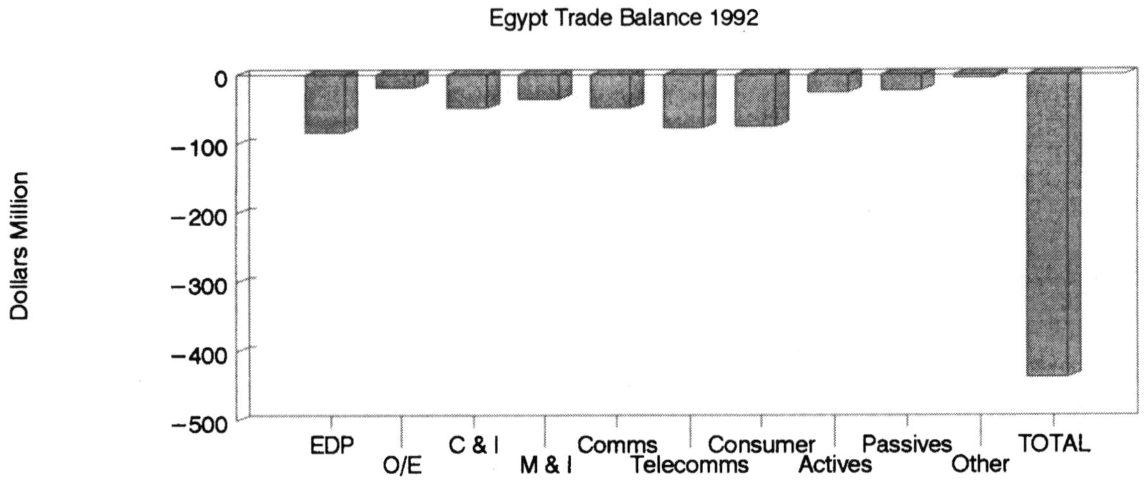

EGYPT – MARKETS

Table No.4.2.2

EDP		1991	1992	1993	1994	1995	1996	1997	CAAGR %
Total EDP Market	E£M	261	377	415	446	475	503	533	7
	$M	**78**	**113**	**125**	**134**	**143**	**151**	**160**	

OFFICE EQUIPMENT		1991	1992	1993	1994	1995	1996	1997	CAAGR %
Electronic Typewriters	E£M	5	9	8	7	7	7	7	−5
	$M	**2**	**3**	**2**	**2**	**2**	**2**	**2**	
	No.th	*10*	*18*	*17*	*16*	*16*	*17*	*18*	
Electronic Calculators,	E£M	17	16	15	15	15	15	15	−1
	$M	**5**	**5**	**5**	**5**	**5**	**5**	**5**	
	No.th	*1400*	*1400*	*1300*	*1310*	*1320*	*1330*	*1340*	
Cash Registers	E£M	1	2	2	2	2	3	3	8
	$M	**0**	**1**	**1**	**1**	**1**	**1**	**1**	
Photocopiers	E£M	53	48	46	46	47	48	49	0
	$M	**16**	**14**	**14**	**14**	**14**	**14**	**15**	
Total Office Equipment Market	E£M	76	75	71	70	71	73	74	0
	$M	**23**	**23**	**21**	**21**	**21**	**22**	**22**	

CONTROL & INSTRUMENTATION		1991	1992	1993	1994	1995	1996	1997	CAAGR %
Total Control & Instrumentation	E£M	158	167	162	160	164	168	174	1
	$M	**47**	**50**	**49**	**48**	**49**	**50**	**52**	

MEDICAL & INDUSTRIAL		1991	1992	1993	1994	1995	1996	1997	CAAGR %
Total X–Ray & Medical Equipment	E£M	49	58	61	65	70	76	82	7
	$M	**15**	**17**	**18**	**20**	**21**	**23**	**25**	
Total Industrial Equipment	E£M	48	65	68	70	72	74	76	3
	$M	**14**	**20**	**20**	**21**	**22**	**22**	**23**	
Total Medical & Industrial Market	E£M	97	123	129	135	142	150	158	5
	$M	**29**	**37**	**39**	**41**	**43**	**45**	**47**	

COMMUNICATIONS & RADAR		1991	1992	1993	1994	1995	1996	1997	CAAGR %
Total Communications & Radar	E£M	188	220	231	245	257	270	283	5
	$M	**56**	**66**	**69**	**74**	**77**	**81**	**85**	

TELECOMMUNICATIONS		1991	1992	1993	1994	1995	1996	1997	CAAGR %
Total Telecommunications Market	E£M	316	311	280	274	277	280	283	−2
	$M	**95**	**93**	**84**	**82**	**83**	**84**	**85**	

CONSUMER		1991	1992	1993	1994	1995	1996	1997	CAAGR %
Video Equipment									
Colour Television	BtM	256	239	227	228	235	243	251	1
	$M	**77**	**72**	**68**	**68**	**71**	**73**	**75**	
	No.th	*460*	*438*	*420*	*425*	*430*	*440*	*450*	
Monochrome Television	E£M	5	4	4	4	3	3	3	−6
	$M	**2**	**1**	**1**	**1**	**1**	**1**	**1**	
	No.th	*25*	*20*	*19*	*18*	*17*	*16*	*15*	

EGYPT – MARKETS (cont)

		1991	1992	1993	1994	1995	1996	1997	
Video Recorders	E£M	68	80	75	76	78	82	86	1
	$M	**20**	**24**	**23**	**23**	**23**	**25**	**26**	
	No.th	*70*	*88*	*83*	*85*	*88*	*92*	*95*	
Video Cameras & Camera Combinations	E£M	1	2	3	4	5	6	7	28
	$M	**0**	**1**	**1**	**1**	**2**	**2**	**2**	
	No.th	*0*	*1*	*2*	*2*	*3*	*3*	*4*	
Total Video Equipment	E£M	330	325	309	312	321	334	347	1
	$M	**99**	**98**	**93**	**94**	**96**	**100**	**104**	
Audio Equipment									
Radio Sets	E£M	35	39	40	41	43	45	46	3
	$M	**11**	**12**	**12**	**12**	**13**	**14**	**14**	
	No.th	*900*	*850*	*870*	*890*	*920*	*950*	*970*	
Tape Recorders, Record Players and CD Players	E£M	28	34	35	36	38	39	41	4
	$M	**8**	**10**	**11**	**11**	**11**	**12**	**12**	
	No.th	*185*	*170*	*174*	*178*	*185*	*190*	*200*	
Total Audio Equipment	E£M	63	73	75	77	81	84	87	4
	$M	**19**	**22**	**23**	**23**	**24**	**25**	**26**	
Personal Consumer Equipment									
Electronic Musical Instruments	E£M	4	4	3	3	3	4	4	0
	$M	**1**	**1**	**1**	**1**	**1**	**1**	**1**	
	No.th	*33*	*30*	*26*	*27*	*28*	*29*	*30*	
Electric/Electronic Watches	E£M	71	56	54	55	57	59	61	2
	$M	**21**	**17**	**16**	**17**	**17**	**18**	**18**	
	No.th	*3300*	*2900*	*2800*	*2900*	*3000*	*3100*	*3200*	
Electric/Electronic Clocks	E£M	16	11	10	11	12	12	13	3
	$M	**5**	**3**	**3**	**3**	**4**	**4**	**4**	
	No.th	*900*	*700*	*650*	*700*	*750*	*770*	*800*	
Total Personal Consumer Equipment	E£M	91	71	67	69	72	75	78	2
	$M	**27**	**21**	**20**	**21**	**22**	**23**	**23**	
Total Consumer Market	E£M	484	469	451	458	474	493	512	2
	$M	**145**	**141**	**135**	**138**	**142**	**148**	**154**	

COMPONENTS		**1991**	**1992**	**1993**	**1994**	**1995**	**1996**	**1997**	CAAGR %
Active Components									
Total Active Components	E£M	104	94	89	88	89	93	97	1
	$M	**31**	**28**	**27**	**26**	**27**	**28**	**29**	
Passive Components									
Total Passive Components	E£M	89	87	83	81	83	85	88	0
	$M	**27**	**26**	**25**	**24**	**25**	**26**	**26**	
Other Components									
Total Other Components	E£M	27	29	28	27	28	28	28	-1
	$M	**8**	**9**	**8**	**8**	**8**	**8**	**8**	
Total Components Market	E£M	220	210	200	196	200	206	213	0
	$M	**66**	**63**	**60**	**59**	**60**	**62**	**64**	

		1991	**1992**	**1993**	**1994**	**1995**	**1996**	**1997**	CAAGR
TOTAL MARKET – EGYPT	E£M	1800	1952	1939	1984	2060	2143	2230	3
	$M	**541**	**586**	**582**	**596**	**619**	**644**	**670**	

Notes: 1991 and 1992 are current figures at current exchange rates
1993 to 1997 are forecasts at 1992 constant values & exchange rates (ie. inflation is not included)
CAAGR column indicates true annual average growth between 1992 & 1997

4.3 GREECE

BASIC STATISTICS

Land Area:		131 937 sq km (80% mainland, 20% islands)
Population:		10.4 million (1993)
Civilian Employment:		3.85 million (1990)
of which:	- Agriculture	23%
	- Industry	27%
	- Other	50%
GDP per Capita:		US$7660 (1992)
TV Sets in Use:		2.3 million
Radio Sets in Use:		4.1 million
Cars per 1000 Inhabitants:		168 (1992)
Main Telephone Lines per 1000 Inhabitants:		408 (1992)

LEADING ECONOMIC INDICATORS (% increases)

	Actual				Forecast		
	1988	**1989**	**1990**	**1991**	**1992**	**1993**	**1994**
Growth of GDP (Constant market prices)	4.1	3.5	0.0	1.8	1.5	1.0	1.5
Consumer Price Index	13.5	13.8	20.5	19.4	15.9	14.5	10.0
Industrial Production	5.2	1.4	-2.4	-1.4	-1.2	0.0	1.5

Overview

Greece is located at the south-eastern tip of Europe and borders with Albania, the former Yugoslavia and Bulgaria to the north, Turkey and the Aegean Sea to the east, and the Mediterranean and Ionian Seas to the south and west. On 1 January 1981, Greece became a full member of the European Community, having been an associate member for 20 years before that. The EC is by far Greece's main trading partner. Greece is also a member of the Council of Europe, NATO and GATT

Greece has had a stable parliamentary government since 1974, when the seven year period of military rule ended. Between 1974 and 1981 Greece was governed by the centre-right New Democracy Party, but in 1981, the Pahellenic Socialist Movement (PASOK), swept to power.

The chronic problems of stagflation and external imbalance are largely the result of the fiscal profligacy of the Socialists (PASOK), who governed the country from 1981 to 1989. The government deficit expanded rapidly, reaching more than 18% of GDP in 1989, and it topped 20% in 1990. This is by far the largest relative deficit in the OECD area, being broadly double even Italy's.

In terms of GDP per head, Greece was the second poorest EC country when it joined in 1981 and is now the poorest having been overtaken by Portugal in 1991.

The breakup of Yugoslavia, across Greece's northern border, has blocked the main trade route to northern Europe. Problems over the recognition of Macedonia and the threat of immigration highlight the uncertainties created by the Yugoslav conflict.

GDP Growth
GDP growth slowed from 1.8% in 1991 to 1.5% in 1992, with the increase in agricultural output down from 10% to 1%. Real private consumption growth was little changed at 1.3%, public consumption picked up a little from the 0.7% cut in 1991, and total fixed investment swung from a 2% fall to a 3% rise. Private investment picked up slightly after a 6% drop in 1991, while public investment (30% of the total) continued to grow at an annual rate of about 10%.

Overall GDP growth is unlikely to exceed 1% in 1993. This compares with negative EC growth of around 0.5%, but a recovery in 1994 should see Greece more convergent with the rest of Europe. The medium term forecasts for growth are no better than for the rest of Europe. This could remain the case at least until inflation is brought down to a lower more sustainable level. Economic policy is firmly tied to achieving convergence with the rest of the EC in the medium term, but stabilisation measures have been needed in recent years to head off a series of incipient crises. The price of EC support has been the adoption of stringent and far reaching deregulation and tough measures to tackle fiscal problems.

Government Policy
Following the collapse of east European communism the two democratic parties in Greece, the New Democracy (ND) and the Panhellenic Socialist Movement (PASOK) consolidated their positions. After eight years' rule, the Panhellenic Socialist Movement suffered a heavy defeat in the general election of June 1989. Its loss of support reflected a mixture of state corruption scandals, failure to deliver promised Socialist reforms, and the continued weakness of the economy. The then second largest party, the conservative New Democracy, joined forces with the communist led Left Alliance to form an interim government, the main purpose of which was to bring corruption charges against ministers of the previous socialist government. The New Democracy continued to govern after the election in April 1990 with a very small majority until October 1993 when they lost power again to the Socialists. However, the new government has committed itself to continue the tight monetary policy and to reinforce the disinflation process started by the conservatives.

An EC loan of ECU 2.2 billion agreed in February 1991 was conditional. ECU 1 billion was paid almost immediately, but the payment of a further two tranches of ECU 600 million each in 1992 and 1993 was to depend upon the progress made on a number of economic objectives, including the reduction of the current account deficit to 3% of GDP and slowing of inflation to 8%. Payment of the 1993 tranche was delayed until the government presented a satisfactory plan for medium-term convergence with the economic performance of EC partners.

Prices & Inflation
High inflation and a severe fiscal imbalance remain the key problems. Anti-inflation policies have brought the inflation rate down from its peak of 22% in 1986 but progress has been slow and intermittent. The annual inflation rate fell from over 18% in early 1992 to a 14.4% low in December, but accelerated to 16.4% in May 1993 . The government's target of less than 10% by the end of 1993 was not achieved. This, however, appears to be a precondition for taking the drachma into the ERM by 1994 at the latest. The government granted a public sector pay award of only 4% and was hoping that this would strongly influence the private

sector wage round. After several nationwide strikes, however, the biggest labour federation won a 13.4% rise for 1993. Influenced by the slack in demand, inflation is expected to fall to around 10% in 1994.

Industrial Outlook
The transformation from a rural to industrial economy has slowed in the past decade. The share of agricultural output in GDP was still as high as 9% in 1991, compared with an EC average of only about 3%, and some 12% in 1980. Agriculture still accounts for over 24% of total employment, compared with 7% for EC as a whole. The performance of the manufacturing sector, which accounts for 27% of employment compared with an EC average of 23%, will continue to be hampered by the predominance of small units specialising in traditional, low technology activities. Family control, an inadequate education system and a brain drain have all impeded the spread of modern business techniques, and Greek manufacturers will therefore continue to find it difficult to make headway against EC competition in the Single European Market. The government has pushed ahead with its privatisation programme with two cement companies and two shipyards already sold and two state refineries being privatised.

The industrial recession bottomed out in 1992 with a 4% rise in production achieved by the fourth quarter. However, order books remain depressed and consumer confidence remains at its weakest level since the early 1980s. The civil war in the former Yugoslavia has underlined the disadvantages of Greece's peripheral location within the EC which takes almost two-thirds of Greek exports. Growth in industrial production is forecast to be zero in 1993 and then to show a modest growth of 1.5% in 1994.

Labour Market
Unemployment rose in Greece in the early 1980s as the world recession caused many Greeks previously working abroad to return home. Since then the number of people employed has remained virtually unchanged with the unemployment rate being around 8%. This is forecast to rise to about 9% in the mid-1990s before increased growth rates cause it to drop slightly towards the end of the decade.

Trade & Balance of Payments
The current account deficit declined from US$3.6 billion (5.5% of GDP) in 1990 to US$1.5 billion (2.2% of GDP) in 1991 as the surplus on invisibles increased, while the trade deficit was virtually unchanged. In 1992, however, another increase in the invisibles surplus, to US$11.9 billion, was not enough to offset the deterioration in the trade deficit to US$13.9 billion, and the overall deficit on current account widened to US$2 billion. An improvement has been evident in 1993. In the five months to April the current account deficit narrowed to an annualized rate of US$1.2 billion.

Although export growth is likely to be constrained by the relatively weak upturn in European markets, import growth is now expected to surge as economic recovery proceeds. Much of the impetus to growth will come from the construction sector and infrastructure projects, which will have low import dependence. The current account deficit could therefore continue to improve, and the privatisation programme may lead to capital inflows from abroad.

ELECTRONICS INDUSTRY

STRUCTURE

Number of Electrical/Electronics Companies: 185 (1988)

Main Production Activities: Telecommunications Equipment, computer peripherals and parts, passive components (mainly switches, inductors & connectors)

Major Companies:

Alpha	Netcom
Elco - Vayonis	Nitelco
Elinda	Pitsos
Elmet	Radio A Konassidis
Fourlis Bros Corp.	Radio Athinai
Intracom	Siemens Tele Industrie
Klimatechniki	Vierex

Electronics/Electrical Industry Employment: 12 130 (1988)

The electronics industry in Greece is still small, total production in 1992 being around US$400 million. Exports are minimal at US$84 million in 1992 of which US$64 was to the EC. The major production is telecommunications equipment which accounts for over 30% of the total. Production of computer equipment has been growing and reached an estimated US$52 million in 1992, a real growth after allowing for inflation of 35%. In the consumer electronics sector, production of colour television sets has been decreasing in favour of cheaper imported sets.

The Greek electronics industry suffers from a poor infrastructure, lack of effective standardisation procedures, inadequate research and development facilities, and poor links between industry and the universities.

One of the main electronics manufacturing companies is Siemens, which manufactures televisions, telephone equipment and professional equipment. Another is the electrical appliance manufacturer Radio Athinia.

Telecommunications services in Greece are provided by OTE, the Hellenic Telecommunications organisation which was founded as a public utility company in 1949. OTE's largest branch is telephony but it also provides telex, telegraphy, data and radio communications, and leased circuit services. In 1988, OTE formulated a five-year development programme which is aimed at the modernisation and improvement of the telecommunications network, through the accelerated large-scale introduction of digital exchanges and digital transmission systems and the introduction of new services including teleinformatics, mobile and satellite communications and ISDN.

Greece is clearly under-equipped in information technology in comparison with the other countries of the EU. A survey by the European Commission in 1987 estimated the Greek market for computers, software and services at 0.3% of the EC's data processing market, whereas the population of Greece represents 3% of the European population. Whilst the market has been growing at a faster pace than the rest of the EC at US$372 million, excluding software and services, in 1994 it is still only 0.6% of the EU.

TRADE WITH THE EUROPEAN COMMUNITY

Values in current US$ millions

	Imports to the EC from Greece			Exports from the EC to Greece		
Product Group	**1990**	**1991**	**1992**	**1990**	**1991**	**1992**
EDP	5	12	15	114	134	142
Office Equipment	0	0	1	17	20	16
Control & Instruments	8	6	8	72	59	87
Medical	1	1	1	24	38	34
Industrial	0	0	0	14	17	20
Communications	29	3	6	49	33	46
Telecommunications	4	6	18	83	139	131
Consumer-Video	2	1	1	106	71	82
Consumer-Audio	0	0	0	8	10	23
Consumer-Personal	0	0	0	11	11	12
Active Components	1	1	1	13	13	12
Passive Components	10	13	13	25	47	53
Other Components	0	12	0	16	56	36
TOTAL	60	55	64	552	648	694

Quantity in thousands of units

Product Group	Imports to the EC from Greece			Exports from the EC to Greece		
	1990	**1991**	**1992**	**1990**	**1991**	**1992**
Colour Television	3	1	1	262	176	199
Monochrome Television	0	0	0	15	7	6
Video Cassette Recorders	0	0	0	1	22	24

INDUSTRY DEVELOPMENTS

- Finanziaria Telefonica has won a twenty year development licence for cellular telephone systems worth US$20 million.
- Redifon SPT has obtained a £2.5 million pact from the Greek MOD for a radio communications system.
- Thomson-CSF approach radar station and telecommunications system is now in service at Athens airport.
- Greece has obtained EEC funding for half of its ECU 292 million telecommunications network development programme.

LIST OF SOURCES

Financial Times
Bank Economic Country Reports
Country Report & Country Profile — EIU
OECD Economic Survey
OECD Economic Outlook
Euromonitor
International Telecom Statistics — Siemens
World Automotive Statistics — SMMT
Predicasts F&S Index International
Electronic Times
Electronic News
Electronic Business
Electronic Engineering Times
Athens Chamber of Commerce & Industry
Greek Embassy
National Statistical Service of Greece
Business & Finance
Exterior Commerce of Greece
Financial Directory of Greek Companies — ICAP

GREECE – PRODUCTION

Table No.4.3.1

EDP		1991	1992	1993	1994	CAAGR %
Computers	DrM	1600	2500	3000	3600	20
	$M	**9**	**13**	**16**	**19**	
Peripherals	DrM	3000	4500	5400	6500	20
	$M	**16**	**24**	**28**	**34**	
Accessories & Parts	DrM	2000	3000	3600	4300	20
	$M	**11**	**16**	**19**	**23**	
Total EDP Production	DrM	6600	10000	12000	14400	20
	$M	**36**	**52**	**63**	**76**	

OFFICE EQUIPMENT		1991	1992	1993	1994	CAAGR %
Electronic Typewriters	DrM	1200	1100	1000	900	–10
	$M	**7**	**6**	**5**	**5**	
	No.th	*36*	36	36	36	
Electronic Cash Registers	DrM	0	4000	3900	4000	0
	$M	**0**	**21**	**20**	**21**	
	No.th	*0*	50	52	55	
Photocopiers	DrM	4000	5000	4800	5000	0
	$M	**22**	**26**	**25**	**26**	
	No.th	*21*	*25*	*25*	*26*	
Total Office Equipment Production	DrM	5200	10100	9700	9900	–1
	$M	**29**	**53**	**51**	**52**	

CONTROL & INSTRUMENTATION		1991	1992	1993	1994	CAAGR %
Industrial & Process Control	DrM	300	350	400	450	13
	$M	**2**	**2**	**2**	**2**	
Instrumentation	DrM	2000	2500	2800	3200	13
	$M	**11**	**13**	**15**	**17**	
Accessories & Parts	DrM	600	800	1000	1200	22
	$M	**3**	**4**	**5**	**6**	
Total Control & Instrumentation	DrM	2900	3650	4200	4850	15
	$M	**16**	**19**	**22**	**25**	

MEDICAL & INDUSTRIAL		1991	1992	1993	1994	CAAGR %
X–Ray & Medical Equipment						
Total X–Ray & Medical Equipment	DrM	250	260	275	310	9
	$M	**1**	**1**	**1**	**2**	
Total Industrial Equipment	DrM	1000	1100	1150	1200	4
	$M	**5**	**6**	**6**	**6**	
Total Medical & Industrial Production	DrM	1250	1360	1425	1510	5
	$M	**7**	**7**	**7**	**8**	

GREECE – PRODUCTION (cont)

COMMUNICATIONS & RADAR		**1991**	**1992**	**1993**	**1994**	CAAGR %
Total Communications & Radar	DrM	5500	7000	7700	8600	11
	$M	**30**	**37**	**40**	**45**	

TELECOMMUNICATIONS		**1991**	**1992**	**1993**	**1994**	CAAGR %
Total Telecommunications Production	DrM	22000	23000	21850	22000	−2
	$M	**121**	**121**	**115**	**115**	

CONSUMER		**1991**	**1992**	**1993**	**1994**	
Video Equipment						
Colour Television	DrM	1720	1750	1800	1900	4
	$M	**9**	**9**	**9**	**10**	
	No.th	*30*	*30*	*31*	*33*	
Monochrome Television	DrM	25	60	50	40	−18
	$M	**0**	**0**	**0**	**0**	
	No.th	*2*	*4*	*4*	*3*	
Video Recorders	DrM	5000	5200	5300	5500	3
	$M	**27**	**27**	**28**	**29**	
	No.th	*110*	*115*	*120*	*125*	
Total Video Equipment	DrM	6745	7010	7150	7440	3
	$M	**37**	**37**	**38**	**39**	
Audio Equipment						
Mains Radios & Combinations	DrM	6000	6500	6615	6800	2
	$M	**33**	**34**	**35**	**36**	
	No.th	*950*	*1100*	*1150*	*1200*	
Tape Recorders	DrM	1400	1400	1450	1500	4
	$M	**8**	**7**	**8**	**8**	
	No.th	*150*	*190*	*195*	*200*	
Record Players	DrM	10	40	35	30	−13
	$M	**0**	**0**	**0**	**0**	
	No.th	*1*	*3*	*3*	*3*	
Total Audio Equipment	NZ$	7410	7940	8100	8330	2
	$M	**41**	**42**	**42**	**44**	
Personal Consumer Equipment						
Electric/Electronic Watches	DrM	300	200	210	220	5
	$M	**2**	**1**	**1**	**1**	
	No.th	*160*	*100*	*100*	*100*	
Total Personal Consumer Equipment	DrM	300	200	210	220	5
	$M	**2**	**1**	**1**	**1**	
Total Consumer Production	DrM	14455	15150	15460	15990	3
	$M	**79**	**79**	**81**	**84**	

GREECE – PRODUCTION (cont)

COMPONENTS		1991	1992	1993	1994	CAAGR %
Active Components						
Valves & Tubes	DrM	50	60	65	70	8
	$M	**0**	**0**	**0**	**0**	
Discrete Semiconductors	DrM	200	400	420	450	6
	$M	**1**	**2**	**2**	**2**	
ICs & Other Microcircuits	DrM	400	600	700	800	15
	$M	**2**	**3**	**4**	**4**	
Total Active Components	DrM	650	1060	1185	1320	12
	$M	**4**	**6**	**6**	**7**	
Passive Components						
Other Fixed Capacitors	DrM	30	30	31	32	3
	$M	**0**	**0**	**0**	**0**	
Fixed Resistors	DrM	50	60	62	64	3
	$M	**0**	**0**	**0**	**0**	
Variable Resistors	DrM	60	100	105	110	5
	$M	**0**	**1**	**1**	**1**	
Connectors	DrM	200	300	350	400	15
	$M	**1**	**2**	**2**	**2**	
Small Transformers, Chokes, Coils and Other Inductors	DrM	1060	1710	1750	1800	3
	$M	**6**	**9**	**9**	**9**	
Relays	DrM	80	160	170	180	6
	$M	**0**	**1**	**1**	**1**	
Switches	DrM	150	150	150	150	0
	$M	**1**	**1**	**1**	**1**	
Printed Circuit Boards	DrM	550	500	500	530	3
	$M	**3**	**3**	**3**	**3**	
Total Passive Components	DrM	2180	3010	3118	3266	4
	$M	**12**	**16**	**16**	**17**	
Other Components						
Amplifiers	DrM	150	150	150	150	0
	$M	**1**	**1**	**1**	**1**	
Aerials	DrM	100	110	120	130	9
	$M	**1**	**1**	**1**	**1**	
Unrecorded Media	DrM	130	150	160	170	6
	$M	**1**	**1**	**1**	**1**	
Accessories & Parts for Consumer Equipment	DrM	350	300	300	310	2
	$M	**2**	**2**	**2**	**2**	
Total Other Components	DrM	730	710	730	760	3
	$M	**4**	**4**	**4**	**4**	
Total Components Production	DrM	3560	4780	5033	5346	6
	$M	**20**	**25**	**26**	**28**	

GREECE – PRODUCTION (cont)

		1991	1992	1993	1994	CAAGR %
TOTAL PRODUCTION – GREECE	DrM	61465	75040	77368	82596	5
	$M	**337**	**394**	**406**	**433**	

Notes: 1991 and 1992 are current figures at current exchange rates
1993 & 1994 are forecasts at 1992 constant values & exchange rates (ie inflation is not included)
CAAGR column indicates true annual average growth between 1992 & 1994

Greece Trade Balance 1992

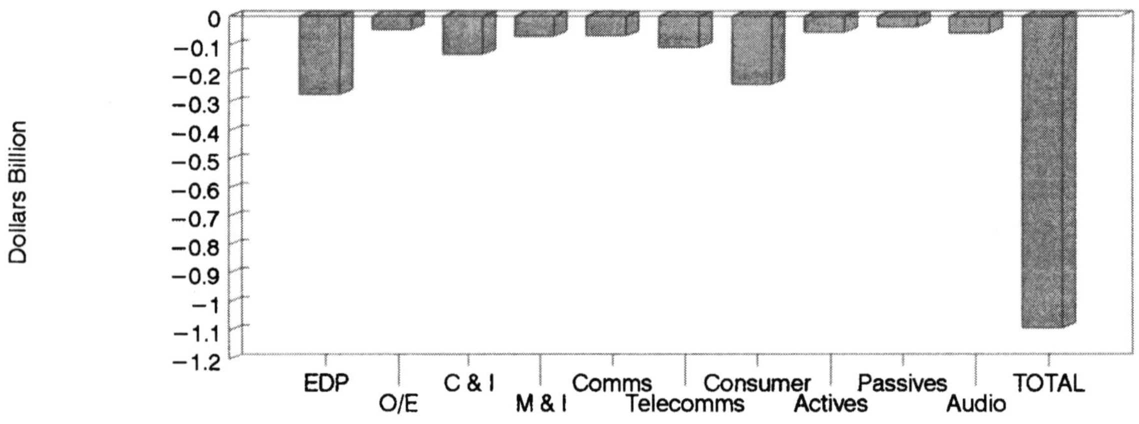

GREECE – MARKETS

Table No.4.3.2

EDP		1991	1992	1993	1994	1995	1996	1997	CAAGR %
Computers	DrM	13756	12617	13200	14300	15900	17800	20000	10
	$M	**75**	**66**	**69**	**75**	**83**	**93**	**105**	
Peripherals	DrM	25795	28734	29500	30700	32500	35000	38000	6
	$M	**141**	**151**	**155**	**161**	**171**	**184**	**199**	
Accessories & Parts	DrM	13111	21066	22800	25780	29460	32800	36000	11
	$M	**72**	**111**	**120**	**135**	**155**	**172**	**189**	
Total EDP Market	DrM	52662	62417	65500	70780	77860	85600	94000	9
	$M	**289**	**327**	**344**	**371**	**408**	**449**	**493**	

OFFICE EQUIPMENT		1991	1992	1993	1994	1995	1996	1997	CAAGR %
Electronic Typewriters	DrM	1510	1421	1350	1280	1210	1150	1100	–5
	$M	**8**	**7**	**7**	**7**	**6**	**6**	**6**	
	No.th	*46*	*47*	*48*	*49*	*50*	*51*	*52*	
Electronic Calculators	DrM	1668	1091	1150	1140	1120	1110	1100	0
	$M	**9**	**6**	**6**	**6**	**6**	**6**	**6**	
	No.th	*1090*	*727*	*800*	*820*	*840*	*860*	*880*	
Electronic Cash Registers	DrM	6794	6723	6600	6500	6400	6300	6200	–2
	$M	**37**	**35**	**35**	**34**	**34**	**33**	**33**	
	No.th	*94*	*90*	*89*	*88*	*87*	*86*	*85*	
Photocopiers	DrM	7880	9646	9200	9000	8870	8700	8500	–2
	$M	**43**	**51**	**48**	**47**	**47**	**46**	**45**	
	No.th	*32*	*36*	*37*	*39*	*40*	*41*	*42*	
Total Office Equipment Market	DrM	17852	18881	18300	17920	17600	17260	16900	–2
	$M	**98**	**99**	**96**	**94**	**92**	**91**	**89**	

CONTROL & INSTRUMENTATION		1991	1992	1993	1994	1995	1996	1997	CAAGR %
Industrial & Process Control	DrM	4118	4819	5200	5700	6300	7000	7800	10
	$M	**23**	**25**	**27**	**30**	**33**	**37**	**41**	
Instrumentation	DrM	14725	19507	19600	20000	20800	21700	22500	3
	$M	**81**	**102**	**103**	**105**	**109**	**114**	**118**	
Accessories & Parts	DrM	2915	4569	4700	5000	5400	5800	6200	6
	$M	**16**	**24**	**25**	**26**	**28**	**30**	**33**	
Total Control & Instrumentation	DrM	21758	28895	29500	30700	32500	34500	36500	5
	$M	**119**	**152**	**155**	**161**	**171**	**181**	**192**	

MEDICAL & INDUSTRIAL		1991	1992	1993	1994	1995	1996	1997	CAAGR %
X–Ray & Medical Equipment									
Total X–Ray & Medical Equipment	DrM	8032	9519	10470	11500	12700	13900	15300	10
	$M	**44**	**50**	**55**	**60**	**67**	**73**	**80**	
Total Industrial Equipment	DrM	3875	5101	5350	5680	6020	6380	6760	6
	$M	**21**	**27**	**28**	**30**	**32**	**33**	**35**	
Total Medical & Industrial Market	DrM	11907	14620	15820	17180	18720	20280	22060	9
	$M	**65**	**77**	**83**	**90**	**98**	**106**	**116**	

GREECE – MARKETS (cont)

COMMUNICATIONS & RADAR		1991	1992	1993	1994	1995	1996	1997	CAAGR %
Total Communications & Radar	DrM	15409	20044	22000	24000	26300	28800	31600	10
	$M	**85**	**105**	**115**	**126**	**138**	**151**	**166**	

TELECOMMUNICATIONS		1991	1992	1993	1994	1995	1996	1997	CAAGR %
Total Telecommunications Market	DrM	40527	43469	43900	44800	45700	46600	47500	2
	$M	**222**	**228**	**230**	**235**	**240**	**244**	**249**	

CONSUMER		1991	1992	1993	1994	1995	1996	1997	CAAGR %
Video Equipment									
Colour Television	DrM	20179	21751	21900	22500	23700	24900	26000	4
	$M	**111**	**114**	**115**	**118**	**124**	**131**	**136**	
	No.th	*367*	*423*	*425*	*440*	*470*	*500*	*520*	
Monochrome Television	DrM	1228	1052	900	800	730	670	600	−11
	$M	**7**	**6**	**5**	**4**	**4**	**4**	**3**	
	No.th	*93*	*75*	*70*	*65*	*60*	*55*	*50*	
Video Recorders	DrM	7534	8382	8300	8500	8900	9300	9800	3
	$M	**41**	**44**	**44**	**45**	**47**	**49**	**51**	
	No.th	*146*	*161*	*160*	*165*	*170*	*180*	*190*	
Video Cameras & Camera Combinations	DrM	1057	2025	2400	2700	3000	3200	3400	11
	$M	**6**	**11**	**13**	**14**	**16**	**17**	**18**	
	No.th	*8*	*16*	*20*	*22*	*24*	*26*	*28*	
Total Video Equipment	DrM	29998	33210	33500	34500	36330	38070	39800	4
	$M	**165**	**174**	**176**	**181**	**191**	**200**	**209**	
Audio Equipment									
Portable Radios & Combinations	DrM	2513	2945	2930	2950	3100	3200	3300	2
	$M	**14**	**15**	**15**	**15**	**16**	**17**	**17**	
	No.th	*1159*	*973*	*980*	*1000*	*1050*	*1100*	*1150*	
Mains Radios & Combinations	DrM	7076	8257	8350	8500	8800	9200	9600	3
	$M	**39**	**43**	**44**	**45**	**46**	**48**	**50**	
	No.th	*1052*	*1208*	*1200*	*1220*	*1250*	*1300*	*1350*	
Car Radios & Combinations	DrM	1619	1684	1650	1670	1750	1850	1950	3
	$M	**9**	**9**	**9**	**9**	**9**	**10**	**10**	
	No.th	*152*	*167*	*160*	*162*	*170*	*175*	*180*	
Tape Recorders	DrM	2260	2659	2700	2750	2820	2900	3000	2
	$M	**12**	**14**	**14**	**14**	**15**	**15**	**16**	
	No.th	*338*	*363*	*370*	*380*	*390*	*400*	*410*	
Record Players	DrM	275	207	200	190	180	170	160	−5
	$M	**2**	**1**	**1**	**1**	**1**	**1**	**1**	
	No.th	*19*	*16*	*15*	*14*	*13*	*12*	*11*	
Compact Disc Players	DrM	569	864	950	1050	1150	1200	1250	8
	$M	**3**	**5**	**5**	**6**	**6**	**6**	**7**	
	No.th	*27*	*33*	*37*	*41*	*45*	*48*	*50*	
Total Audio Equipment	DrM	14312	16616	16780	17110	17800	18520	19260	3
	$M	**79**	**87**	**88**	**90**	**93**	**97**	**101**	

GREECE – MARKETS (cont)

Personal Consumer Equipment									
Electronic Musical Instruments	DrM	1296	895	890	900	920	940	960	1
	$M	**7**	**5**	**5**	**5**	**5**	**5**	**5**	
Electric/Electronic Watches	DrM	6555	8501	8420	8600	9000	9400	9800	3
	$M	**36**	**45**	**44**	**45**	**47**	**49**	**51**	
	No.th	*3590*	*4070*	*4000*	*4100*	*4300*	*4500*	*4700*	
Electric/Electronic Clocks	DrM	1365	1224	1200	1220	1260	1300	1340	2
	$M	**7**	**6**	**6**	**6**	**7**	**7**	**7**	
	No.th	*1527*	*1304*	*1300*	*1320*	*1350*	*1380*	*1400*	
Electronic Flashlights	DrM	87	53	52	53	54	55	56	1
	$M	**0**	**0**	**0**	**0**	**0**	**0**	**0**	
Total Personal Consumer Equipment	DrM	9303	10673	10562	10773	11234	11695	12156	3
	$M	**51**	**56**	**55**	**57**	**59**	**61**	**64**	
Total Consumer Market	DrM	53613	60499	60842	62383	65364	68285	71216	3
	$M	**294**	**317**	**319**	**327**	**343**	**358**	**374**	

COMPONENTS		**1991**	**1992**	**1993**	**1994**	**1995**	**1996**	**1997**	CAAGR %
Active Components									
Colour TV Tubes	DrM	512	527	540	570	600	630	660	5
	$M	**3**	**3**	**3**	**3**	**3**	**3**	**3**	
Monochrome TV Tubes	DrM	6	15	13	10	9	8	7	−14
	$M	**0**	**0**	**0**	**0**	**0**	**0**	**0**	
Other Valves & Tubes	DrM	832	233	240	250	260	270	280	4
	$M	**5**	**1**	**1**	**1**	**1**	**1**	**1**	
Discrete Semiconductors	DrM	1651	1935	1970	2050	2150	2250	2350	4
	$M	**9**	**10**	**10**	**11**	**11**	**12**	**12**	
ICs & Other Microcircuits	DrM	8715	9082	9380	10000	10900	11850	12850	7
	$M	**48**	**48**	**49**	**52**	**57**	**62**	**67**	
Total Active Components	DrM	11716	11792	12143	12880	13919	15008	16147	6
	$M	**64**	**62**	**64**	**68**	**73**	**79**	**85**	
Passive Components									
Electrolytic Capacitors	DrM	84	137	140	145	150	155	160	3
	$M	**0**	**1**	**1**	**1**	**1**	**1**	**1**	
Other Fixed Capacitors	DrM	1773	2184	2230	2320	2420	2520	2620	4
	$M	**10**	**11**	**12**	**12**	**13**	**13**	**14**	
Variable Capacitors	DrM	225	436	400	410	420	430	440	0
	$M	**1**	**2**	**2**	**2**	**2**	**2**	**2**	
Fixed Resistors	DrM	355	399	410	420	430	440	450	2
	$M	**2**	**2**	**2**	**2**	**2**	**2**	**2**	
Variable Resistors	DrM	286	267	270	280	290	300	310	3
	$M	**2**	**1**	**1**	**1**	**2**	**2**	**2**	
Connectors	DrM	1361	1465	1500	1560	1620	1680	1740	4
	$M	**7**	**8**	**8**	**8**	**8**	**9**	**9**	
Relays	DrM	453	471	475	490	500	510	520	2
	$M	**2**	**2**	**2**	**3**	**3**	**3**	**3**	
Switches	DrM	1055	1476	1500	1540	1570	1600	1630	2
	$M	**6**	**8**	**8**	**8**	**8**	**8**	**9**	

GREECE – MARKETS (cont)

		1991	1992	1993	1994	1995	1996	1997	
Small Transformers, Chokes, Coils and Other Inductors	DrM	1086	1158	1170	1200	1230	1260	1290	2
	$M	**6**	**6**	**6**	**6**	**6**	**7**	**7**	
Printed Circuit Boards	DrM	1924	2282	2385	2535	2700	2900	3100	6
	$M	**11**	**12**	**13**	**13**	**14**	**15**	**16**	
Total Passive Components	DrM	8602	10275	10480	10900	11330	11795	12260	4
	$M	**47**	**54**	**55**	**57**	**59**	**62**	**64**	
Other Components									
Microphones	DrM	179	259	260	265	270	275	280	2
	$M	**1**	**1**	**1**	**1**	**1**	**1**	**1**	
Loudspeakers	DrM	2325	3135	3200	3300	3400	3500	3600	3
	$M	**13**	**16**	**17**	**17**	**18**	**18**	**19**	
Amplifiers	DrM	681	844	860	890	920	950	980	3
	$M	**4**	**4**	**5**	**5**	**5**	**5**	**5**	
Aerials	DrM	1193	1522	1550	1600	1650	1700	1750	3
	$M	**7**	**8**	**8**	**8**	**9**	**9**	**9**	
Unrecorded Media	DrM	4528	5193	5300	5500	5700	5900	6100	3
	$M	**25**	**27**	**28**	**29**	**30**	**31**	**32**	
Accessories & Parts for Consumer Equipment	DrM	703	977	1000	1020	1040	1060	1080	2
	$M	**4**	**5**	**5**	**5**	**5**	**6**	**6**	
Total Other Components	DrM	9609	11930	12170	12575	12980	13385	13790	3
	$M	**53**	**63**	**64**	**66**	**68**	**70**	**72**	
Total Components Market	DrM	29927	33997	34793	36355	38229	40188	42197	4
	$M	**164**	**178**	**183**	**191**	**201**	**211**	**221**	

		1991	1992	1993	1994	1995	1996	1997	CAAGR
TOTAL MARKET – GREECE	DrM	243655	282822	290655	304118	322273	341513	361973	5
	$M	**1337**	**1484**	**1525**	**1596**	**1691**	**1792**	**1899**	

Notes: 1991 and 1992 are current figures at current exchange rates
1993 to 1997 are forecasts at 1992 constant values & exchange rates (ie. inflation is not included)
CAAGR column indicates true annual average growth between 1992 & 1997

4.4 MEXICO

BASIC STATISTICS

Land Area:		1.97 million sq km
Population:		88 million (1991)
Civilian Employment:		28 million (1990)
of which:	- Agriculture	23%
	- Manufacturing	27%
	- Other	50%
GDP per Capita:		US$3700 (1992 estimate)
TV Sets in Use:		5.6 million
Radio Sets in Use:		16.3 million
Cars per 1000 Inhabitants:		80 (1992)
Main Telephone Lines per 1000 Inhabitants:		71 (1992)

LEADING ECONOMIC INDICATORS (% increases)

	Actual				Forecast		
	1988	**1989**	**1990**	**1991**	**1992**	**1993**	**1994**
Growth of GDP							
(Constant Market Price)	3.5	3.2	4.4	3.6	2.2	0.4	3.0
Consumer Price Index	52	20	30	20	15	9.0	8.0
Manufacturing Output	3.0	6.5	5.2	2.0	1.8	0.0	na

Overview

Mexico lies between the southern border of the United States and the continent of South America. The country is successfully pursuing market oriented economic reforms and is preparing for entry to the North American Free Trade Area (NAFTA), together with the United States and Canada. the transition to a market economy is proving difficult, as the opening of domestic markets to foreign trade and economic recovery has resulted in large trade and current account deficits. Mexico's government is stable and recent legislation has considerably liberalised foreign investment potential. The country also has a low cost, easily trainable work-force. Mexico's currency is the peso.

GDP Growth

The rate of economic expansion slackened in 1992, with 2.2% rise in Gross Domestic Product the lowest growth rate for four years. The deceleration reflected the impact of high interest rates, a weaker world economy and adjustment difficulties associated with the process of economic restructuring. To maintain the country's attractiveness to foreign investors and cut inflation to single figures, tight fiscal and monetary policies are being pursued, at the expense of economic growth. GDP growth is forecast to be less than 1% in 1993 but then to accelerate to around 3% in 1994.

Government Policy

The ruling party, PRI, has been in power since 1929. It recorded poor results in certain of the state governorship elections held in 1992. However, the PRI emerged as the clear victor in three state elections held in the early part of 1993. In power since December 1988, President Carlos Salinas de Gortari is constitutionally barred from standing for office in the 1994 presidential elections. The Social Development minister and the Mayor of Mexico City are possible successors to Salinas. The economic stability pact, agreed by government, business and unions in 1987 and since renewed several times, was further extended in late October 1992. The pact will now run at least until the end of 1993. The agreement, which covers wages, salaries, administered prices and the exchange rate, is an important instrument for combating inflation.

The Salinas government has been particularly successful in consolidating public sector finances, systematically cutting the size of the public sector, removing subsidies and raising administrative charges to cover costs.

Prices & Inflation

The budget deficit, the main factor behind the high inflation rates of earlier years, has been eliminated. A budget surplus, equivalent to 0.5% of GDP, was recorded in 1992. This figure excludes the substantial revenues from the privatisation of state owned enterprises, which are not automatically recorded as state revenue but instead are allocated to a reserve fund in the Finance Ministry. Inflation, which at its peak in 1987 exceeded 160%, declined to 12% in 1992 and 10.4% in March of 1993. As the year progresses, it should fall below 10% for the first time since 1972, and is forecast to reduce still further in 1994.

Industrial Outlook

Accounting for roughly a quarter of GDP and 60% of exports, the industrial sector is undergoing profound structural changes. Output increased by just 1.8% in 1992, the lowest increase over the past six years. Following the liberalisation of imports in recent years and the real appreciation of the peso, various industries, such as textiles and metals, have been facing strong foreign competition. Construction activity, however, increased at an above average rate of 7.8%, while exports from the maquiladora (in-bond) industry, one of the most important sources of foreign exchange, increased by 16% to US$4.8 billion.

Mexico is the sixth most important producer of petroleum and is ranked seventh internationally in terms of petroleum reserves. The US is by far the most important purchaser of Mexican crude, followed by Spain and Japan. The proportion of exports accounted for by petroleum fell from almost 70% in 1981 to some 30% in 1992.

Agriculture contributes around 8% of GDP. The sector continues to suffer from structural problems and will require substantial investment if productivity is to be raised. In 1992 output fell by 1.5%. The number of state-run companies was reduced from 1155 in 1982 to some 225 by the end of 1992. Eighteen commercial banks nationalised in 1982, have been re-privatised, generating revenues of over US$12 billion. The sale in 1993 of a further 37 state-owned organisations, including several television stations and various ports will mark the conclusion of the privatisation process.

Labour Market

In 1990 the workforce was estimated to number some 28 million. That this is a relatively small proportion of the total of the population reflects the fact that a large percentage is under 15 years of age and, more importantly, that only about half of those in the appropriate age group actively enter the formal labour market. Official sources put a figure of 2.8% on unemployment in the 16 main urban centres in the first nine months of 1990. If those working for less than 15 hours a week had been counted as unemployed, the figure would have been 6.5%. Depressed economic growth in 1992 and 1993 may have increased this figure.

Trade & Balance of Payments
Following the opening of domestic markets to foreign trade and the exchange rate induced fall in the real cost of imports, the trade deficit has been widening. Despite the economic slowdown in 1992, imports rose by 26% to US$48.1 billion, while exports remained virtually unchanged at US$278.5 billion. The trade deficit widened from US$11.1 billion in 1991 to US$20.6 billion, while the current account deficit, which has been deteriorating since 1988, exceeded US$22 billion, a significant increase from the US$13.3 billion recorded in 1991. Sizeable inflows of foreign capital, largely in the form of foreign direct or portfolio investment, have financed the current account deficit. The current account deficit is forecast to increase further, possibly to as high as US$25 billion in 1994.

ELECTRONICS INDUSTRY

STRUCTURE

Main Production Activities: Consumer Electronics, Computers

Major Companies:

America Tel SA de CV
Control y Aseguramiento SA de CV
Electropartes de Matamoros SA de CV
Hewlett-Packard de Mexico
IBM de Mexico SA de CV
Industria Mexicana Unisys
Industrios Unidas SA
Kodak Mexicana SA de CV
Logica Digital
Mitel de Mexico
Motorola de Mexico
Nashua de Mexico
NCR Industrial de Mexico
Olivetti Mexicana
Olympia de Mexico
Partes de Television de Reynosa SA de CV
Philips Mexicana SA de CV
Printaform
RCA Components SA de CV
Siemens
Teleindustria Ericsson SA de CV
Wang de Mexico
XEROX Mexicana

There were 71 683 electronics industry employees in 1989.

The major electronic companies operating in Mexico are still American-owned, these are IBM de Mexico, Xerox Mexicana, and Kodak Mexicana.

The majority of electronics production in Mexico is carried out through the Maquiladora programme. The term Maquiladora refers to an assembly plant operating in Mexico territory under special customs treatment and liberal foreign investment regulations. Maquiladoras import into Mexico, duty free, machinery, equipment, parts, raw materials and other components. Finished or semi finished products are exported back to their country of origin or to a third country.

The total number of Maquiladora plants in operation at the end of September 1993 was 2182, up 4.8% over a year earlier. Employment was 552 078, representing more than 20% of total employment in the Mexican manufacturing sector, up 7.4% from 1992. Whilst most plants are clustered along the border the number in the interior has increased by more than 40% in the past 4 years. The trend is expected to continue as new plants are built in Yucatan, Jalisco and Quintana Roo. Around a third of the Maquiladora plants and a similar proportion of employees are engaged in electrical and electronic production.

Most Maquiladoras are foreign owned, principally by companies from the US but also from Korea, Japan, Taiwan, Hong Kong, Sweden, France and Canada.

Japan is the third largest foreign investor in Mexico, after the US and Germany, with about 6.5% of total investment. Traditionally, Japan has been the leading investor in the electronics sectors, but only in recent years has investment in the Maquiladora sector started to take off. Most of these plants were set up by US subsidiaries of Japanese corporations that had production facilities in the US before establishing Maquiladoras.

Companies which have made investments in the Maquiladora industry in recent years include:

- Matsushita (colour television equipment)
- Sony (colour television components and sound equipment)
- Mitsui
- Canon
- Toshiba
- Pioneer
- Casio

Mexico is an important producer of consumer goods for the US market, in particular, around 6.5 million colour television sets are produced each year for sale to the US.

Many companies are looking at Mexico's interior as an alternative site for in-bond assembly (maquiladora) operations away from the crowded US-Mexico border area. Japanese companies are making strategic investments in Mexico's maquiladora industry with the primary aim of increasing competitiveness in the US market. They see considerable scope for expansion in the northern border zone, yet they are moving cautiously in setting up these projects. Like their US counterparts, Japanese maquiladora are affected by inefficiencies in the sector, including inadequate infrastructure, poor raw materials and high labour turnover.

After months of negotiation between the government and Mexico's computer industry, the Commerce Secretariat announced in April 1990 two decrees opening the computer industry to foreign competition. One decree lifts the requirement that computer imports receive prior government authorisation, while the other is designed to encourage local sourcing in the newly competitive market. Together, the decrees have caused multinational companies with computer companies operating in Mexico to rethink local strategies and view Mexico within the context of their global operations. The new decrees specifically provide for the lifting of all import permits for finished computers, parts and accessories.

The opening has boosted Mexico's rapidly growing informatics market, the second largest in Latin America, after Brazil. Imports increased by over 100% between 1990 and 1992 but are forecast to slow in 1993 with the downturn in the economy. Production is increasing at a modest rate of around 5% per annum.

TRADE WITH THE EUROPEAN COMMUNITY

Values in current US$ millions

Product Group	Imports to the EC from Mexico 1989	1990	1991	1992	Exports from the EC to Mexico 1989	1990	1991	1992
EDP	21	39	60	85	26	36	71	64
Office Equipment	0	1	13	9	4	6	6	11
Control & Instruments	9	3	7	4	33	61	106	129
Medical	0	1	4	6	9	13	12	17
Industrial	2	1	0	2	0	13	5	7
Communications	1	1	1	1	20	30	35	63
Telecommunications	1	2	2	2	90	210	127	208
Consumer-Video	1	0	1	0	0	1	1	1
Consumer-Audio	1	5	18	16	16	1	2	4
Consumer-Personal	2	2	4	3	2	2	2	3
Active Components	15	17	16	12	7	7	9	11
Passive Components	5	9	17	17	1	23	55	60
Other Components	1	1	16	17	2	6	29	15
TOTAL	59	82	159	174	210	409	460	593

Quantity in thousands of units

Product Group	Imports to the EC from Mexico 1991	1992	Exports from the EC to Mexico 1991	1992
Colour Television	0	0	0	1
Monochrome Television	0	0	0	0
Video Cassette Recorders	2	0	0	2

INDUSTRY DEVELOPMENTS

- TEMIC is building a microsystem plant in Cuautia.
- Samsung Electronics is to expand the production capacity of its Mexican plant from 1 million colour televisions per year to 1.2 million.
- Siemens has opened a new factory for automotive electronics in the Guadalay area.
- Grupo Financiers Banamex Acio is to form a joint venture to build fibre optic networks in Mexico City, Monterray and Guadbedgaru.
- Thompson-CSF & Alenia win against competion to upgrade Mexican ATC.
- Sanyo Electric is to spend US$25M to build two 10 000 sq ft electronics parts plants in Tyuana.
- Northern Telecom has announced it is to relocate its portable telephone production

plant from Santa Clare in California to Monterray in early 1994. It is expected to create 900 jobs in Monterray.

- Nintendo, a subsidiary of Nintendo of Japan will lay-off 10% of its 1300 workers in Redmond US and move assembly to Mexico.

- Hewlett Packard's Mexican subsidiary recently opened a new 178 000 sq ft headquarters in the Santa Fe development project west of Mexico City.

- The Toshiba Corporation in March 1994 announced plans to move production of small television sets from Tenessee to a new plant in Mexico. The Mexican plant, by reducing labour costs, will allow Toshiba to compete with production from South Korea.

- Del Computer Corporation announced plans to open a plant in Mexico in mid-1993 to perform customer configaration on a range of desk top computers. The plant at Toluca will support operations at its Dell Computer de Mexico unit.

- Fifteen Korean firms will set up plants in Northern Mexico in the secold half of 1993 to produce electronic components. The Korean facility will produce around 400 jobs. Foreign capital will be used to construct the plants but Mexican firms will later help the Koreans to develop the factories.

- Key Tronics of Spokane Washington, a manufacturer of computer keyboards, plans to eliminate 100 full-time jobs in Washington and add 100 to its plant at Juarex Mexico. Most manufacturing will shift to Juarex and to plants in Ireland, Malaya and New Mexico

LIST OF SOURCES

Financial Times
Bank Economic Country Reports
Country Report & Country Profile — EIU
International Telecom Statistics — Siemens
World Automotive Statistics — SMMT
Predicasts F&S Index International
Electronic Times
Electronic News
Electronic Business
Electronic Engineering Times
Mexican Traders
Business Latin America
Euromonitor
CANIECE
US Imports & Exports — US Department of Commerce

MEXICO – PRODUCTION

Table No.4.4.1

EDP		1991	1992	1993	1994	CAAGR %
Complete Systems	PsB	2160	2170	2294	2449	
	$M	**720**	**700**	**740**	**790**	6
Peripherals	PsB	1380	1705	1767	1829	
	$M	**460**	**550**	**570**	**590**	4
Accessories & Parts	PsB	150	186	217	248	
	$M	**50**	**60**	**70**	**80**	15
Total EDP Production	PsB	3690	4061	4278	4526	
	$M	**1230**	**1310**	**1380**	**1460**	6

OFFICE EQUIPMENT		1991	1992	1993	1994	CAAGR %
Electronic Typewriters	PsB	60	47	43	40	
	$M	**20**	**15**	**14**	**13**	-7
	No.th	*190*	140	130	120	
Electronic Calculators	PsB	6	6	6	9	
	$M	**2**	**2**	**2**	**3**	22
	No.th	*450*	430	430	600	
Photocopiers	PsB	132	264	298	326	
	$M	**44**	**85**	**96**	**105**	11
	No.th	*30*	*52*	*60*	*70*	
Total Office Equipment Production	PsB	198	316	347	375	9
	$M	**66**	**102**	**112**	**121**	

CONTROL & INSTRUMENTATION		1991	1992	1993	1994	CAAGR %
Total Control & Instrumentation	PsB	1020	1085	1116	1178	
	$M	**340**	**350**	**360**	**380**	4

MEDICAL & INDUSTRIAL		1991	1992	1993	1994	CAAGR %
Medical Equipment						
Radiation Equipment	PsB	153	186	192	202	
	$M	**51**	**60**	**62**	**65**	4
Other Electromedical Equipment (including Hearing Aids)	PsB	750	884	930	1011	
	$M	**250**	**285**	**300**	**326**	7
Total X-Ray & Medical Equipment	PsB	903	1070	1122	1212	
	$M	**301**	**345**	**362**	**391**	6
Industrial Equipment						
Railway & Traffic Signalling Equipment	PsB	9	19	19	22	
	$M	**3**	**6**	**6**	**7**	8
Security & Fire Alarms & Other Signalling Equipment	PsB	207	192	195	214	
	$M	**69**	**62**	**63**	**69**	5
Total Industrial Equipment	PsB	216	211	214	236	
	$M	**72**	**68**	**69**	**76**	6
Total Medical & Industrial Production	PsB	1119	1280	1336	1448	
	$M	**373**	**413**	**431**	**467**	6

MEXICO – PRODUCTION (cont)

COMMUNICATIONS & RADAR		1991	1992	1993	1994	CAAGR %
Total Communications & Radar	PsB	780	775	791	837	
	$M	**260**	**250**	**255**	**270**	4

TELECOMMUNICATIONS		1991	1992	1993	1994	CAAGR %
Total Telecommunications Production	PsB	825	930	939	967	
	$M	**275**	**300**	**303**	**312**	2

CONSUMER		1991	1992	1993	1994	
Video Equipment						
Colour Television	PsB	2850	3875	3953	4185	
	$M	**950**	**1250**	**1275**	**1350**	4
	No.th	*4900*	*6400*	*6500*	*6900*	
Total Video Equipment	PsB	2850	3875	3953	4185	
	$M	**950**	**1250**	**1275**	**1350**	4
Audio Equipment						
Radios (including Car Radios)	PsB	1920	1922	1860	1922	
	$M	**640**	**620**	**600**	**620**	0
	No.th	*6300*	*5200*	*5000*	*5200*	
Tape Recorders & Decks	PsB	480	685	651	682	
	$M	**160**	**221**	**210**	**220**	0
	No.th	*2700*	*2900*	*2800*	*2900*	
CD Players	PsB	138	530	558	620	
	$M	**46**	**171**	**180**	**200**	8
	No.th	*360*	*520*	*600*	*700*	
Total Audio Equipment	PsB	2538	3137	3069	3224	1
	$M	**846**	**1012**	**990**	**1040**	
Personal Consumer Equipment						
Electric/Electronic Watches	PsB	54	62	65	68	
	$M	**18**	**20**	**21**	**22**	5
	No.th	*1500*	*1600*	*1700*	*1800*	
Electric/Electronic Clocks	PsB	6	6	6	6	
	$M	**2**	**2**	**2**	**2**	0
	No.th	*350*	*340*	*330*	*320*	
Total Personal Consumer Equipment	PsB	60	68	71	74	
	$M	**20**	**22**	**23**	**24**	4
Total Consumer Production	PsB	5448	7080	7093	7483	
	$M	**1816**	**2284**	**2288**	**2414**	3

MEXICO – PRODUCTION (cont)

COMPONENTS		1991	1992	1993	1994	CAAGR %
Active Components						
Other Valves & Tubes	PsB	9	9	9	9	
	$M	**3**	**3**	**3**	**3**	0
Diodes	PsB	255	310	316	326	
	$M	**85**	**100**	**102**	**105**	2
Transistors	PsB	255	248	236	233	
	$M	**85**	**80**	**76**	**75**	-3
Discrete Semiconductors	PsB	177	177	180	186	
	$M	**59**	**57**	**58**	**60**	3
ICs & Other Microcircuits	PsB	1560	1643	1693	1876	
	$M	**520**	**530**	**546**	**605**	7
Total Active Components	PsB	2256	2387	2434	2629	
	$M	**752**	**770**	**785**	**848**	5
Passive Components						
Capacitors	PsB	645	698	729	769	
	$M	**215**	**225**	**235**	**248**	5
Variable Capacitors	PsB	9	12	12	12	
	$M	**3**	**4**	**4**	**4**	0
Fixed Resistors	PsB	72	78	81	84	
	$M	**24**	**25**	**26**	**27**	4
Variable Resistors	PsB	150	155	155	155	
	$M	**50**	**50**	**50**	**50**	0
Connectors	PsB	570	605	620	636	
	$M	**190**	**195**	**200**	**205**	3
Small Transformers, Chokes Coils and Other Inductors	PsB	474	521	558	605	
	$M	**158**	**168**	**180**	**195**	8
Relays	PsB	120	118	115	112	
	$M	**40**	**38**	**37**	**36**	-3
Switches	PsB	390	471	465	465	
	$M	**130**	**152**	**150**	**150**	-1
Printed Circuit Boards	PsB	270	310	350	403	
	$M	**90**	**100**	**113**	**130**	14
Total Passive Components	PsB	2700	2967	3085	3240	
	$M	**900**	**957**	**995**	**1045**	4
Other Components						
Microphones	PsB	48	56	56	56	
	$M	**16**	**18**	**18**	**18**	0
Loudspeakers	PsB	126	149	155	161	
	$M	**42**	**48**	**50**	**52**	4
Amplifiers	PsB	114	109	105	102	
	$M	**38**	**35**	**34**	**33**	-3
Aerials	PsB	30	31	31	34	
	$M	**10**	**10**	**10**	**11**	5
Unrecorded Media	PsB	330	310	326	341	
	$M	**110**	**100**	**105**	**110**	5
Accessories & Parts for Consumer Equipment	PsB	1050	1116	1147	1209	
	$M	**350**	**360**	**370**	**390**	4
Total Other Components	PsB	1698	1770	1820	1903	
	$M	**566**	**571**	**587**	**614**	4

MEXICO – PRODUCTION (cont)

		1991	1992	1993	1994	
Total Components Production	PsB	6654	7124	7338	7772	
	$M	**2218**	**2298**	**2367**	**2507**	4

		1991	1992	1993	1994	CAAGR %
TOTAL PRODUCTION – MEXICO	PsB	19734	22652	23238	24586	
	$M	**6578**	**7307**	**7496**	**7931**	4

Notes: 1991 and 1992 are current figures at current exchange rates
1993 & 1994 are forecasts at 1992 constant values & exchange rates (ie inflation is not included)
Current estimates & forecasts are made in US$ with local values shown for guidance only
CAAGR column indicates true annual average growth between 1992 & 1994

Mexico Trade Balance 1992

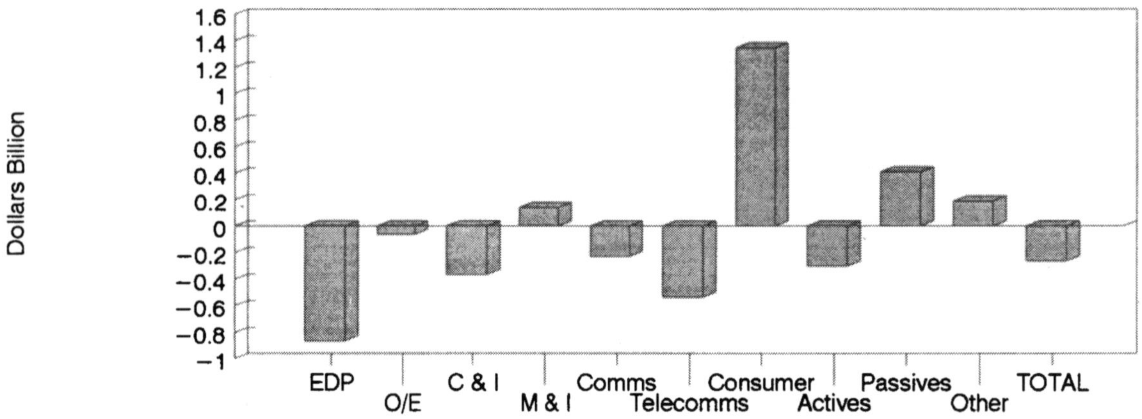

MEXICO – MARKETS

Table No.4.4.2

EDP		1991	1992	1993	1994	1995	1996	1997	CAAGR %
Complete Systems	PsB	2097	2570	2542	2697	2883	3100	3317	
	$M	**699**	**829**	**820**	**870**	**930**	**1000**	**1070**	5
Peripherals	PsB	2514	3091	3069	3162	3286	3410	3596	
	$M	**838**	**997**	**990**	**1020**	**1060**	**1100**	**1160**	3
Accessories & Parts	PsB	828	1076	1054	1085	1178	1271	1333	
	$M	**276**	**347**	**340**	**350**	**380**	**410**	**430**	4
Total EDP Market	PsB	5439	6736	6665	6944	7347	7781	8246	
	$M	**1813**	**2173**	**2150**	**2240**	**2370**	**2510**	**2660**	4

OFFICE EQUIPMENT		1991	1992	1993	1994	1995	1996	1997	CAAGR %
Electronic Typewriters	PsB	108	102	99	96	93	90	87	
	$M	**36**	**33**	**32**	**31**	**30**	**29**	**28**	−3
	No.th	*310*	*273*	*274*	*275*	*276*	*277*	*278*	
Electronic Calculators	PsB	30	31	31	34	37	37	40	
	$M	**10**	**10**	**10**	**11**	**12**	**12**	**13**	5
	No.th	*2000*	*1900*	*2000*	*2100*	*2200*	*2300*	*2400*	
Electronic Cash Registers	PsB	66	136	136	140	140	140	143	
	$M	**22**	**44**	**44**	**45**	**45**	**45**	**46**	1
	No.th	*35*	*60*	*63*	*65*	*67*	*69*	*71*	
Dictation Machines	PsB	33	43	43	43	47	47	47	
	$M	**11**	**14**	**14**	**14**	**15**	**15**	**15**	1
	No.th	*150*	*180*	*183*	*186*	*189*	*190*	*191*	
Photocopiers	PsB	180	198	211	223	239	251	260	
	$M	**60**	**64**	**68**	**72**	**77**	**81**	**84**	6
	No.th	*60*	*62*	*70*	*77*	*85*	*95*	*105*	
Total Office Equipment Market	PsB	417	512	521	536	555	564	577	
	$M	**139**	**165**	**168**	**173**	**179**	**182**	**186**	2

CONTROL & INSTRUMENTATION		1991	1992	1993	1994	1995	1996	1997	CAAGR %
Total Control & Instrumentation	PsB	1917	2207	2139	2248	2427	2604	2728	
	$M	**639**	**712**	**690**	**725**	**783**	**840**	**880**	4

MEDICAL & INDUSTRIAL		1991	1992	1993	1994	1995	1996	1997	CAAGR %
Medical Equipment									
Medical X–Ray Equipment	PsB	96	93	93	96	99	102	105	
	$M	**32**	**30**	**30**	**31**	**32**	**33**	**34**	3
Industrial X–Ray Equipment	PsB	24	22	22	22	25	25	28	
	$M	**8**	**7**	**7**	**7**	**8**	**8**	**9**	5
Radiation Equipment	PsB	21	22	22	25	25	28	28	
	$M	**7**	**7**	**7**	**8**	**8**	**9**	**9**	5
Other Electromedical Equipment (including Hearing Aids)	PsB	417	474	487	543	605	673	735	
	$M	**139**	**153**	**157**	**175**	**195**	**217**	**237**	9
Total X–Ray & Medical Equipment	PsB	558	611	623	685	753	828	896	
	$M	**186**	**197**	**201**	**221**	**243**	**267**	**289**	8

MEXICO – MARKETS (cont)

		1991	1992	1993	1994	1995	1996	1997	
Industrial Equipment									
Railway & Traffic Signalling Equipment	PsB	39	31	31	34	37	40	43	
	$M	**13**	**10**	**10**	**11**	**12**	**13**	**14**	7
Security & Fire Alarms & Other Signalling Equipment	PsB	120	140	155	171	192	214	236	
	$M	**40**	**45**	**50**	**55**	**62**	**69**	**76**	11
Induction & Dielectric Equipment	PsB	21	56	53	56	59	62	62	
	$M	**7**	**18**	**17**	**18**	**19**	**20**	**20**	2
Total Industrial Equipment	PsB	180	226	239	260	288	316	341	
	$M	**60**	**73**	**77**	**84**	**93**	**102**	**110**	9
Total Medical & Industrial Market	PsB	738	837	862	946	1042	1144	1237	
	$M	**246**	**270**	**278**	**305**	**336**	**369**	**399**	8

COMMUNICATIONS & RADAR		**1991**	**1992**	**1993**	**1994**	**1995**	**1996**	**1997**	CAAGR %
Total Communications & Radar	PsB	1080	1488	1519	1593	1705	1829	1938	
	$M	**360**	**480**	**490**	**514**	**550**	**590**	**625**	5

TELECOMMUNICATIONS		**1991**	**1992**	**1993**	**1994**	**1995**	**1996**	**1997**	CAAGR %
Total Telecommunications Market	PsB	2439	2582	2505	2530	2579	2635	2685	
	$M	**813**	**833**	**808**	**816**	**832**	**850**	**866**	1

CONSUMER		**1991**	**1992**	**1993**	**1994**	**1995**	**1996**	**1997**	CAAGR %
Video Equipment									
Colour Television	PsB	750	1066	1023	1054	1104	1141	1184	
	$M	**250**	**344**	**330**	**340**	**356**	**368**	**382**	2
	No.th	*997*	*1271*	*1250*	*1300*	*1350*	*1400*	*1450*	
Monochrome Television	PsB	57	43	40	37	37	34	34	
	$M	**19**	**14**	**13**	**12**	**12**	**11**	**11**	−5
	No.th	*230*	*180*	*170*	*160*	*155*	*150*	*150*	
Video Recorders	PsB	417	524	487	496	527	546	564	
	$M	**139**	**169**	**157**	**160**	**170**	**176**	**182**	1
	No.th	*440*	*520*	*490*	*500*	*530*	*550*	*570*	
Video Cameras & Camera Combinations	PsB	87	121	118	130	143	155	167	
	$M	**29**	**39**	**38**	**42**	**46**	**50**	**54**	7
	No.th	*50*	*67*	*65*	*70*	*75*	*80*	*85*	
Total Video Equipment	PsB	1311	1755	1668	1717	1810	1876	1950	
	$M	**437**	**566**	**538**	**554**	**584**	**605**	**629**	2
Audio Equipment									
Radios (inc Car Radios)	PsB	681	828	809	831	859	887	911	
	$M	**227**	**267**	**261**	**268**	**277**	**286**	**294**	2
	No.th	*2608*	*2996*	*2900*	*3000*	*3100*	*3200*	*3300*	
Tape Recorders & Decks	PsB	57	62	59	62	65	65	68	
	$M	**19**	**20**	**19**	**20**	**21**	**21**	**22**	2
	No.th	*379*	*442*	*420*	*440*	*460*	*470*	*480*	
Compact Disc Players	PsB	54	62	65	68	74	78	81	
	$M	**18**	**20**	**21**	**22**	**24**	**25**	**26**	5
	No.th	*184*	*199*	*210*	*220*	*230*	*240*	*250*	
Total Audio Equipment	PsB	792	952	933	961	998	1029	1060	
	$M	**264**	**307**	**301**	**310**	**322**	**332**	**342**	2

MEXICO – MARKETS (cont)

		1991	1992	1993	1994	1995	1996	1997	
Personal Consumer Equipment									
Electronic Musical Instruments	PsB	33	50	50	50	53	53	53	
	$M	11	16	16	16	17	17	17	1
Electric/Electronic Watches	PsB	108	109	112	118	124	130	133	
	$M	36	35	36	38	40	42	43	4
	No.th	*6000*	*5800*	*6000*	*6300*	*6600*	*6900*	*7100*	
Electric/Electronic Clocks	PsB	33	40	40	43	43	43	47	
	$M	11	13	13	14	14	14	15	3
	No.th	*1750*	*1840*	*1870*	*1900*	*1930*	*1960*	*2000*	
Total Personal Equipment	PsB	174	198	202	211	220	226	233	
	$M	58	64	65	68	71	73	75	3
Total Consumer Market	PsB	2277	2905	2802	2889	3029	3131	3243	
	$M	759	937	904	932	977	1010	1046	2

COMPONENTS		1991	1992	1993	1994	1995	1996	1997	CAAGR %
Active Components									
Colour Television Tubes	PsB	834	1110	1132	1209	1271	1333	1395	
	$M	278	358	365	390	410	430	450	5
Monochrome Television Tubes	PsB	3	3	3	3	3	3	3	
	$M	1	1	1	1	1	1	1	0
Other Valves & Tubes	PsB	66	74	74	78	81	81	84	
	$M	22	24	24	25	26	26	27	2
Diodes	PsB	150	174	177	183	192	198	205	
	$M	50	56	57	59	62	64	66	3
Transistors	PsB	117	140	143	146	152	155	158	
	$M	39	45	46	47	49	50	51	3
Other Discrete Semiconductors	PsB	78	99	99	102	109	115	118	
	$M	26	32	32	33	35	37	38	3
ICs & Other Microcircuits	PsB	1452	1721	1758	1869	2071	2226	2356	
	$M	484	555	567	603	668	718	760	6
Total Active Components	PsB	2700	3320	3385	3590	3878	4111	4318	
	$M	900	1071	1092	1158	1251	1326	1393	5
Passive Components									
Capacitors	PsB	195	220	217	226	239	248	257	
	$M	65	71	70	73	77	80	83	3
Variable Capacitors	PsB	12	12	12	12	12	12	12	
	$M	4	4	4	4	4	4	4	0
Fixed Resistors	PsB	75	71	68	71	74	78	78	
	$M	25	23	22	23	24	25	25	2
Variable Resistors	PsB	27	34	31	31	34	34	37	
	$M	9	11	10	10	11	11	12	2
Connectors	PsB	414	499	496	515	543	574	595	
	$M	138	161	160	166	175	185	192	4
Small Transformers, Chokes Coils,Ballast & Other Inductors	PsB	102	112	109	112	115	118	121	
	$M	34	36	35	36	37	38	39	2
Relays	PsB	81	93	90	93	96	99	102	
	$M	27	30	29	30	31	32	33	2

MEXICO – MARKETS (cont)

		1991	1992	1993	1994	1995	1996	1997	
Switches	PsB	81	93	90	93	96	99	102	
	$M	**27**	**30**	**29**	**30**	**31**	**32**	**33**	2
Printed Circuit Boards	PsB	465	564	552	580	608	648	679	
	$M	**155**	**182**	**178**	**187**	**196**	**209**	**219**	4
Total Passive Components	PsB	1452	1699	1665	1733	1817	1910	1984	
	$M	**484**	**548**	**537**	**559**	**586**	**616**	**640**	3
Other Components									
Microphones	PsB	15	19	19	19	22	22	22	
	$M	**5**	**6**	**6**	**6**	**7**	**7**	**7**	3
Loudspeakers	PsB	75	102	99	102	109	115	118	
	$M	**25**	**33**	**32**	**33**	**35**	**37**	**38**	3
Amplifiers	PsB	78	87	84	87	90	93	96	
	$M	**26**	**28**	**27**	**28**	**29**	**30**	**31**	2
Aerials	PsB	90	96	99	102	109	115	118	
	$M	**30**	**31**	**32**	**33**	**35**	**37**	**38**	4
Unrecorded Media	PsB	237	335	322	341	363	388	403	
	$M	**79**	**108**	**104**	**110**	**117**	**125**	**130**	4
Accessories & Parts for Consumer	PsB	486	546	536	558	580	601	626	
Equipment	**$M**	**162**	**176**	**173**	**180**	**187**	**194**	**202**	3
Total Other Components	PsB	981	1184	1159	1209	1271	1333	1383	
	$M	**327**	**382**	**374**	**390**	**410**	**430**	**446**	3
Total Components Market	PsB	5133	6203	6209	6532	6966	7353	7685	
	$M	**1711**	**2001**	**2003**	**2107**	**2247**	**2372**	**2479**	4

		1991	1992	1993	1994	1995	1996	1997	CAAGR
TOTAL MARKET–MEXICO	PsB	19440	23470	23222	24217	25649	27041	28337	
	$M	**6480**	**7571**	**7491**	**7812**	**8274**	**8723**	**9141**	4

Notes: 1991 and 1992 are current figures at current exchange rates
1993 to 1997 are forecasts at 1992 constant values & exchange rates (ie. inflation is not included)
Current estimates & forecasts are made in US$ with local values shown for guidance only
CAAGR column indicates true annual average growth between 1992 & 1997

4.5 NEW ZEALAND

BASIC STATISTICS

Land Area:		268 112 sq km
Population:		3.43 million (1992)
Civilian Employment:		1.47 million 1992)
of which:	- Agriculture	11%
	- Industry	22%
	- Other	67%
GDP per Capita:		US$12426 (1991)
TV Sets per 1000 Inhabitants:		328 (1989)
Radio Sets per 1000 Inhabitants:		925 (1989)
Cars per 1000 Inhabitants:		459 (1992)
Main Telephone Lines per 1000 Inhabitants:		439 (1992)

LEADING ECONOMIC INDICATORS (% increases)

	Actual				Forecast		
	1988	**1989**	**1990**	**1991**	**1992**	**1993**	**1994**
Growth of GDP (Constant market prices)	2.3	-0.1	1.1	-2.1	2.0	3.0	3.5
Consumer Price Index	6.4	5.7	6.1	2.6	1.0	1.7	1.4

Overview
New Zealand lies in the South Pacific Ocean, some 1600 km to the south-east of Australia.

New Zealand is a small, open economy, still heavily dependent on exports of primary and agricultural products. The major industries in New Zealand concentrate on processing the primary produce of the country. The output can be categorised as meat and animal by-products, dairy products, fish, horticultural products and forestry-based products such as pulp and paper. However, secondary industries have shown steady growth in recent years. They include oil refining, aluminium smelting, carpet manufacturing, vehicle assembly and a wide variety of small-scale specialised manufacturing units.

The last two decades have seen a dramatic fall in living standards as real commodity prices and the terms of trade have worsened. New Zealand has slipped from having one of the highest per capita income levels in the OECD area in the 1950s, to one of the lowest in the early 1990s. The economic crisis in the early 1980s required dramatic structural change and this is still shaping the performance of the economy in the early 1990s. New Zealand was badly hit by the two oil price hikes of the 1970s, and took longer than most other OECD countries to adjust to higher energy prices.

New Zealand is a member of the Commonwealth, the United Nations, the OECD and the Closer Economic Relationship with Australia.

GDP Growth
The New Zealand economy emerged from a long period of recession in the second quarter of 1991 and since then growth has been the strongest for nearly a decade. The main driving force behind this has been a swing in stockbuilding which alone contributed over 2% of

GDP. The other areas of support have been private consumption following a recovery in local confidence and real incomes.

External demand continues to add impetus to the recovery and domestic demand is forecast to remain buoyant in 1994. After under-performing the rest of the OECD area in the 1980s, New Zealand now has the potential to perform much more strongly. A period of balanced growth is now in prospect with low inflation, competitive exports and a balanced external position forming the basis for a prolonged period of growth.

Government Policy

There is a stable two party system, and elections are held once every three years. The Labour party held office in August 1987 to October 1990 period, but suffered a landslide defeat to the opposition National party in the general election held in October 1990.

The National party has pursued an enterprise culture with less reliance on state welfare and subsidies, but suffered a substantial decline in the share of the vote and parliamentary seats in the election held in November 1993. The initial uncertainty following this outcome does not appear to have had any lasting effect on business and consumer confidence.

At the same time as the November election a binding referendum decided to replace the current electoral system by a mixed member proportional system, similar to the one used in Germany, which usually allows minor parties greater representation. The new system will be introduced in the October 1996 general election. It promises to dramatically alter the political structure of New Zealand, with the prospect of a multi-party coalition-type system of government. However, it will not alter the 3 year fixed term of government, which is arguably too short for the implementation of longer-term reform programmes.

Prices & Inflation

The 1980s were characterised by double-digit inflation. Annual inflation peaked in 1987 at 16%. A combination of tight monetary policies, flat domestic demand and industry restructuring has seen inflation fall quicker than expected. The annual inflation rate fell from 5% in the September 1990 quarter to only 2.2% in Sept 1991, and then to 1% in September 1992. The inflation rate may now move back to 2% or less in view of dollar slippage. However, the central bank is continuing to target an inflation rate of 0% to 2% by the end of 1994 and this looks achievable providing the NZ dollar does not fall dramatically.

Industrial Outlook

The structural weaknesses of New Zealand's economy include its heavy reliance on primary products and its over dependence on a handful of large export markets.

The most important areas of manufacture, in terms of total output and employment, include meat and dairy processing, transport equipment, textiles, pulp and paper, metal industries and chemicals. Manufacturing is generally on a small scale, the main exceptions being in pulp and paper, base metals (steel and aluminium) and natural gas-based chemical developments in Taranaki.

Productivity growth was one of the lowest in the OECD area over the 1960-79 period. In the 1960-73 period labour productivity growth reached 1.7% per annum, compared with an average of 4.1% for all industrialised countries. In the following six years it fell at a rate of 1.3% per annum, compared with a rise of 1.4% for the OECD area as a whole. In the 11 years to 1990 labour productivity improved relative to the OECD to 1.5% per annum, just above the average for all industrial countries.

Improved growth in some of New Zealand's major export markets, notably Australia and the US, and also the prospect of more favourable external trading countries resulting from the conclusion of the GATT regulations has created a climate of optimism across most sectors of the economy.

Labour Market
Unemployment was, up to the mid-seventies, extremely low by the standards of most developed countries. However, during the 1980s the number of registered unemployed continued to rise as labour shedding continued in the private and public sectors as a result of economic and public sector reforms as well as a weaker domestic economy. The unemployment rate was 10.3%. The recovery has seen employment growth increase substantially. In the year to September 1993 numbers employed increased by 41 800 compared to an increase of 8800 during the year to September 1992. Numbers unemployed have fallen at a slower rate due to new entrants to the labour force, a rise in labour force participation and fewer New Zealanders departing overseas. The rate of unemployment was 10.4% in 1992 but did reduce to 9.5% in 1993 and is expected to further reduce in the coming years.

Trade & Balance of Payments
The current account has improved from a deficit of NZ$1.7 billion in 1990 to broad balance or small surplus in 1991 and 1992. This has reflected in part the impact of recession on domestic demand (investment fell 16.5% in 1991), but exports have also played an important role. De-stocking and competitive export prices pushed export volume growth substantially higher; in the year to 1992 merchandise exports rose 11.5% in real terms after almost 10% growth in the previous year. Import volume growth recorded 9.7% in 1992 but fell 3.8% in 1991. The reorientation of export markets away from Europe and towards the more buoyant South East Asia area has also been a beneficial experience.

While growth of domestic absorbtion has increased export demand, sustained export demand and further term of trade improvements are expected to protect the balance of payment current account.

ELECTRONICS INDUSTRY

STRUCTURE

There were approximately 150 Electronics/Electrical Companies in 1985.

Main Production Activities: Communications Equipment, Control & Instrumentation, Passive Components, Computer Parts.

Main Electronics Companies:

- Actronic Systems
- Asea Electric
- Aucom Electronics
- Avalon Audio
- AWA NZ
- Brandt
- EMC Electronics
- GPT NZ
- Integrated Technology
- Oscmar Products
- PDL Electronics
- Plessey Business Systems
- Sanyo New Zealand
- Siemens

Fisher & Paykel Electronics
GEC (New Zealand)
Gough Technology
Solid State Equipment
Standard Telephone & Cable (NZ)
Tait Electronics

Electronics Industry Employment: 2686 (1989)

Prior to 1982 consumer electronic goods resembled 66% of the value of electronics production in New Zealand. With some notable exceptions the sector tended to be dominated by local subidiaries of major multinationals and their brands. The main thrust of the 1982 New Zealand Electronics Development Plan was to phase out the production of consumer electronic production in favour of industrial, commercial and communications equipment that was competitive in export markets. The Plan was reviewed in 1986, the major outcome being to remove all import controls by October 1987 and the phasing down of tariffs by 1990. By 1991 the main consumer items still being manufactured were white goods plus a few niche market products.

Strong growth has been enjoyed in agricultural technology, health, commercial/industrial and professional radar and communications equipment. Growth in these areas is attributable to New Zealand's pre-eminance in associated fields such as agriculture.

A need for research and development support is regarded as essential. The criteria of the new Ministry of Science and Technology in 1990 has gone some way to meeting this need.

A major restrictive factor in the development of the electronics industry in New Zealand is the country's distant location from world markets, which increases the relative cost of transport of both finished goods and imported raw materials.

TRADE WITH THE EUROPEAN COMMUNITY

Values in current US$ millions

	Imports to the EC from New Zealand				Exports from the EC to New Zealand			
Product Group	**1989**	**1990**	**1991**	**1992**	**1989**	**1990**	**1991**	**1992**
EDP	1	3	4	3	25	32	30	23
Office Equipment	0	0	0	0	6	3	4	3
Control & Instruments	1	3	3	4	5	18	17	17
Medical	0	0	0	0	3	4	4	3
Industrial	0	0	0	0	1	1	1	1
Communications	0	4	1	1	5	19	12	9
Telecommunications	0	2	3	2	26	43	23	12
Consumer-Video	0	0	0	0	0	7	6	4
Consumer-Audio	0	0	0	0	0	1	1	1
Consumer-Personal	0	0	0	0	0	1	1	1
Active Components	0	0	0	0	0	1	1	1
Passive Components	0	0	1	2	0	0	9	8
Other Components	0	0	3	0	3	2	10	3
TOTAL	2	12	15	12	74	132	119	86

Quantity in thousands of units

Product Group	**Imports to the EC from New Zealand**		**Exports from the EC to new Zealand**	
	1991	**1992**	**1991**	**1992**
Colour Television	0	0	5	8
Monochrome Television	0	0	0	0
Video Cassette Recorders	0	0	16	0

INDUSTRY DEVELOPMENTS

- Electrocomponents purchased Radio Spares Components for £1.2 million cash in 1994
- Broadcast Communications has formed a joint venture with Clear Communications for laying a fibre optic cable at a cost of NZ$40 million.

LIST OF SOURCES

Financial Times
Bank Economic Country Reports
Country Report & Country Profile — EIU
OECD Economic Survey
OECD Economic Outlook
Euromonitor
International Telecom Statistics — Siemens
World Automotive Statistics — SMMT
Far Eastern Economic Review
Predicasts F&S Index International
Electronic Times
Electronic News
Electronic Business
Electronic Engineering Times
New Zealand Business Who's Who
New Zealand Department of Statistics

NEW ZEALAND – PRODUCTION

Table No.4.5.1

EDP		1991	1992	1993	1994	CAAGR %
Complete Systems	NZ$	5	10	15	20	41
	$M	**3**	**5**	**8**	**11**	
Peripherals	NZ$	10	10	25	30	73
	$M	**6**	**5**	**13**	**16**	
Accessories & Parts	NZ$	18	25	30	38	23
	$M	**10**	**13**	**16**	**20**	
Total EDP Production	NZ$	33	45	70	88	40
	$M	**19**	**24**	**38**	**47**	

OFFICE EQUIPMENT		1991	1992	1993	1994	CAAGR %
Total Office Equipment Production	NZ$	0	0	0	0	
	$M	**0**	**0**	**0**	**0**	

CONTROL & INSTRUMENTATION		1991	1992	1993	1994	CAAGR %
Industrial & Process Control	NZ$	57	66	70	80	10
	$M	**33**	**35**	**38**	**43**	
Instrumentation	NZ$	16	23	40	50	47
	$M	**9**	**12**	**22**	**27**	
Accessories & Parts	NZ$	4	8	17	22	66
	$M	**2**	**4**	**9**	**12**	
Total Control & Instrumentation	NZ$	77	97	127	152	25
	$M	**45**	**52**	**68**	**82**	

MEDICAL & INDUSTRIAL		1991	1992	1993	1994	CAAGR %
X–Ray & Medical Equipment						
Other Electromedical Equipment	NZ$	16	18	20	22	11
	$M	**9**	**10**	**11**	**12**	
Total X–Ray & Medical Equipment	NZ$	16	18	20	22	11
	$M	**9**	**10**	**11**	**12**	
Industrial Equipment						
Security & Fire Alarms	NZ$	5	5	4	4	–11
	$M	**3**	**3**	**2**	**2**	
Other Signalling Equipment	NZ$	5	6	7	8	15
	$M	**3**	**3**	**4**	**4**	
Total Industrial Equipment	NZ$	10	11	11	12	4
	$M	**6**	**6**	**6**	**6**	
Total Medical & Industrial Production	NZ$	26	29	31	34	8
	$M	**15**	**16**	**17**	**18**	

COMMUNICATIONS & RADAR		1991	1992	1993	1994	CAAGR %
Radio Communications Equipment	NZ$	43	50	55	63	12
	$M	**25**	**27**	**30**	**34**	
Accessories & Parts	NZ$	2	3	6	7	53
	$M	**1**	**2**	**3**	**4**	
Total Communications & Radar	NZ$	45	53	61	70	15
	$M	**26**	**28**	**33**	**38**	

NEW ZEALAND – PRODUCTION (cont)

TELECOMMUNICATIONS		1991	1992	1993	1994	CAAGR %
Switching Equipment	NZ$	12	17	20	21	11
	$M	**7**	**9**	**11**	**11**	
Data & Text Terminal Equipment including facsimile machines	NZ$	8	8	8	8	0
	$M	**5**	**4**	**4**	**4**	
Transmission Equipment	NZ$	23	24	25	26	4
	$M	**13**	**13**	**13**	**14**	
Telephone Sets	NZ$	12	12	12	12	0
	$M	**7**	**6**	**6**	**6**	
Other Telecommunications Equipment	NZ$	5	8	8	8	0
	$M	**3**	**4**	**4**	**4**	
Accessories & Parts	NZ$	15	14	16	18	13
	$M	**9**	**8**	**9**	**10**	
Total Telecommunications Production	NZ$	75	83	89	93	6
	$M	**43**	**45**	**48**	**50**	

CONSUMER		1991	1992	1993	1994	
Video Equipment						
Colour Television	NZ$	3	3	3	3	0
	$M	**2**	**2**	**2**	**2**	
	No.th	*5*	*5*	*5*	*5*	
Total Video Equipment	NZ$	3	3	3	3	0
	$M	**2**	**2**	**2**	**2**	
Audio Equipment						
Portable Radios & Radio Recorders	NZ$	2	3	3	3	0
	$M	**1**	**2**	**2**	**2**	
	No.th	*40*	*60*	*60*	*62*	
Mains Radios & Combinations	NZ$	1	0	1	0	
	$M	**1**	**0**	**1**	**0**	
	No.th	*8*	*5*	*4*	*4*	
Tape Recorders	NZ$	2	1	1	1	0
	$M	**1**	**1**	**1**	**1**	
	No.th	*25*	*14*	*15*	*16*	
Record Players	NZ$	1	0	0	0	
	$M	**1**	**0**	**0**	**0**	
	No.th	*10*	*0*	*0*	*0*	
Total Audio Equipment	NZ$	6	4	5	4	0
	$M	**3**	**2**	**3**	**2**	
Personal Consumer Equipment						
Total Personal Consumer Equipment	NZ$	0	0	0	0	
	$M	**0**	**0**	**0**	**0**	
Total Consumer Production	NZ$	9	7	8	7	0
	$M	**5**	**4**	**4**	**4**	

COMPONENTS		1991	1992	1993	1994	CAAGR %
Active Components						
Semiconductor	NZ$	9	1	1	1	0
	$M	**5**	**1**	**1**	**1**	
Total Active Components	NZ$	9	1	1	1	0
	$M	**5**	**1**	**1**	**1**	

NEW ZEALAND – PRODUCTION (cont)

Passive Components						
Connectors	NZ$	3	6	7	8	15
	$M	**2**	**3**	**4**	**4**	
Inductors	NZ$	1	1	1	1	0
	$M	**1**	**1**	**1**	**1**	
Printed Circuit Boards	NZ$	14	14	15	16	7
	$M	**8**	**8**	**8**	**9**	
Total Passive Components	NZ$	18	21	23	25	9
	$M	**10**	**11**	**12**	**13**	
Other Components						
Loudspeakers	NZ$	1	0	0	0	
	$M	**1**	**0**	**0**	**0**	
Amplifiers	NZ$	1	0	0	0	
	$M	**1**	**0**	**0**	**0**	
Unrecorded Media	NZ$	3	4	2	2	−29
	$M	**2**	**2**	**1**	**1**	
Accessories & Parts for Consumer Equipment	NZ$	2	3	3	3	0
	$M	**1**	**2**	**2**	**2**	
Total Other Components	NZ$	7	7	5	5	−15
	$M	**4**	**4**	**3**	**3**	
Total Components Production	NZ$	34	29	29	31	3
	$M	**20**	**16**	**16**	**17**	

		1991	1992	1993	1994	CAAGR %
TOTAL PRODUCTION – NEW ZEALAND	NZ$	299	343	415	475	18
	$M	**173**	**184**	**223**	**255**	

Notes: 1991 and 1992 are current figures at current exchange rates
1993 & 1994 are forecasts at 1992 constant values & exchange rates (ie inflation is not included)
CAAGR column indicates true annual average growth between 1992 & 1994

New Zealand Trade Balance 1992

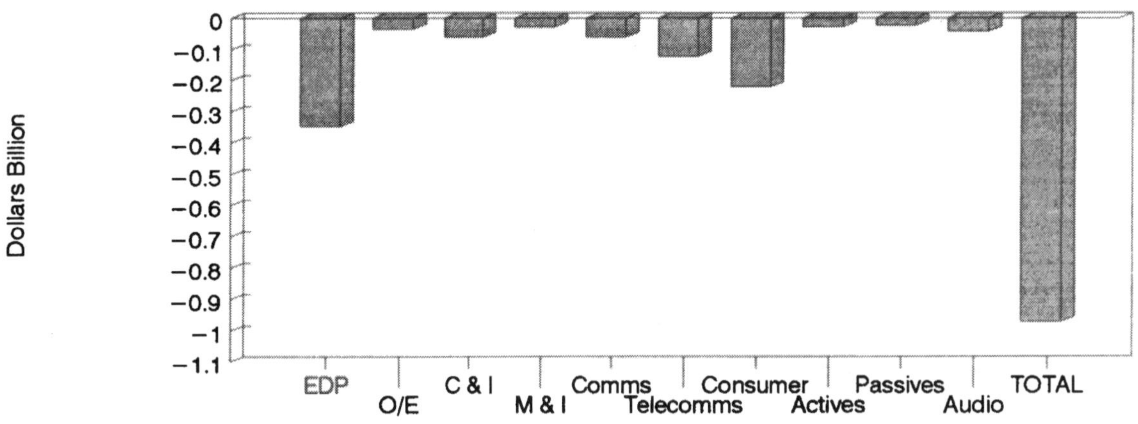

NEW ZEALAND – MARKETS

Table No.4.5.2

EDP		1991	1992	1993	1994	1995	1996	1997	CAAGR %
Complete Systems	NZ$	225	234	240	270	285	305	325	7
	$M	**130**	**126**	**129**	**145**	**153**	**164**	**175**	
Peripherals Storage	NZ$	23	33	33	37	41	44	47	7
	$M	**13**	**18**	**18**	**20**	**22**	**24**	**25**	
Peripherals Input/Output	NZ$	81	94	150	140	145	150	155	11
	$M	**47**	**51**	**81**	**75**	**78**	**81**	**83**	
Peripherals Other	NZ$	83	118	110	125	133	140	150	5
	$M	**48**	**63**	**59**	**67**	**72**	**75**	**81**	
Accessories & Parts	NZ$	167	211	235	257	275	293	311	8
	$M	**97**	**113**	**126**	**138**	**148**	**158**	**167**	
Total EDP Market	NZ$	579	690	768	829	879	932	988	7
	$M	**335**	**371**	**413**	**446**	**473**	**501**	**531**	

OFFICE EQUIPMENT		1991	1992	1993	1994	1995	1996	1997	CAAGR %
Electronic Typewriters	NZ$	14	8	8	8	8	8	7	−3
	$M	**8**	**4**	**4**	**4**	**4**	**4**	**4**	
	No.th	*47*	*32*	*37*	*38*	*39*	*40*	*41*	
Electronic Calculators	NZ$	5	6	7	7	7	7	8	6
	$M	**3**	**3**	**4**	**4**	**4**	**4**	**4**	
	No.th	*436*	*565*	*657*	*670*	*685*	*700*	*720*	
Electronic Cash Registers	NZ$	5	4	5	5	6	6	6	8
	$M	**3**	**2**	**3**	**3**	**3**	**3**	**3**	
	No.th	*6*	*5*	*5*	*5*	*6*	*6*	*6*	
Dictation Equipment	NZ$	2	1	1	1	1	1	1	0
	$M	**1**	**1**	**1**	**1**	**1**	**1**	**1**	
	No.th	*7*	*6*	*5*	*5*	*6*	*6*	*6*	
Photocopiers	NZ$	41	46	52	56	59	62	65	7
	$M	**24**	**25**	**28**	**30**	**32**	**33**	**35**	
	No.th	*9*	*13*	*16*	*17*	*18*	*19*	*20*	
Total Office Equipment Market	NZ$	67	65	73	77	81	84	87	6
	$M	**39**	**35**	**39**	**41**	**44**	**45**	**47**	

CONTROL & INSTRUMENTATION		1991	1992	1993	1994	1995	1996	1997	CAAGR %
Industrial & Process Control	NZ$	71	79	88	95	99	104	110	7
	$M	**41**	**42**	**47**	**51**	**53**	**56**	**59**	
Instrumention	NZ$	69	103	105	110	115	121	126	4
	$M	**40**	**55**	**56**	**59**	**62**	**65**	**68**	
Accessories & Parts	NZ$	25	29	32	34	36	38	40	7
	$M	**14**	**16**	**17**	**18**	**19**	**20**	**22**	
Total Control & Instrumentation	NZ$	165	211	225	239	250	263	276	6
	$M	**95**	**113**	**121**	**128**	**134**	**141**	**148**	

NEW ZEALAND – MARKETS (cont)

MEDICAL & INDUSTRIAL		1991	1992	1993	1994	1995	1996	1997	CAAGR %
X–Ray & Medical Equipment									
X–Ray & Radiation Equipment	NZ$	9	12	14	16	17	19	20	11
	$M	**5**	**6**	**8**	**9**	**9**	**10**	**11**	
Electrocardiographs	NZ$	1	1	1	1	1	1	1	0
	$M	**1**	**1**	**1**	**1**	**1**	**1**	**1**	
Other Electromedical Equipment	NZ$	31	40	51	62	73	85	95	19
	$M	**18**	**22**	**27**	**33**	**39**	**46**	**51**	
Hearing Aids	NZ$	4	5	6	7	8	9	10	15
	$M	**2**	**3**	**3**	**4**	**4**	**5**	**5**	
Total X–Ray & Medical Equipment	NZ$	45	58	72	86	99	114	126	17
	$M	**26**	**31**	**39**	**46**	**53**	**61**	**68**	
Industrial Equipment									CAAGR
Security & Fire Alarms	NZ$	6	8	10	11	11	12	13	10
	$M	**3**	**4**	**5**	**6**	**6**	**6**	**7**	
Other Signalling Equipment	NZ$	10	13	16	18	20	22	24	13
	$M	**6**	**7**	**9**	**10**	**11**	**12**	**13**	
Induction & Dielectric Equipment	NZ$	1	1	2	2	2	2	2	15
	$M	**1**	**1**	**1**	**1**	**1**	**1**	**1**	
Total Industrial Equipment	NZ$	17	22	28	31	33	36	39	12
	$M	**10**	**12**	**15**	**17**	**18**	**19**	**21**	
Total Medical & Industrial Market	NZ$	62	80	100	117	132	150	165	16
	$M	**36**	**43**	**54**	**63**	**71**	**81**	**89**	

COMMUNICATIONS & RADAR		1991	1992	1993	1994	1995	1996	1997	CAAGR %
Radar & Navigational Aids	NZ$	29	27	12	11	11	10	10	–18
	$M	**17**	**15**	**6**	**6**	**6**	**5**	**5**	
Communications Equipment	NZ$	74	130	162	176	187	200	213	10
	$M	**43**	**70**	**87**	**95**	**101**	**108**	**115**	
Accessories & Parts	NZ$	6	10	7	7	7	8	8	–4
	$M	**3**	**5**	**4**	**4**	**4**	**4**	**4**	
Total Communications & Radar	NZ$	109	167	181	194	205	218	231	7
	$M	**63**	**90**	**97**	**104**	**110**	**117**	**124**	

TELECOMMUNICATIONS		1991	1992	1993	1994	1995	1996	1997	CAAGR %
Switching Equipment	NZ$	90	60	62	59	58	57	56	–1
	$M	**52**	**32**	**33**	**32**	**31**	**31**	**30**	
Data & Text Terminal Equipment including facsimile machines	NZ$	28	36	39	40	41	42	43	4
	$M	**16**	**19**	**21**	**22**	**22**	**23**	**23**	
Transmission Equipment	NZ$	52	52	58	55	54	53	52	0
	$M	**30**	**28**	**31**	**30**	**29**	**28**	**28**	
Telephone Sets	NZ$	34	36	45	46	47	48	49	6
	$M	**20**	**19**	**24**	**25**	**25**	**26**	**26**	
Other Telecoms Equipment	NZ$	27	22	13	12	13	14	15	–7
	$M	**16**	**12**	**7**	**6**	**7**	**8**	**8**	
Accessories & Parts	NZ$	41	107	62	61	63	65	67	–9
	$M	**24**	**58**	**33**	**33**	**34**	**35**	**36**	
Total Telecommunications Market	NZ$	272	313	279	273	276	279	282	–2
	$M	**157**	**168**	**150**	**147**	**148**	**150**	**152**	

NEW ZEALAND – MARKETS (cont)

CONSUMER		1991	1992	1993	1994	1995	1996	1997	CAAGR %
Video Equipment									
Colour Television	NZ$	141	159	162	169	176	183	190	4
	$M	**82**	**85**	**87**	**91**	**95**	**98**	**102**	
	No.th	*275*	*297*	*323*	*338*	*353*	*367*	*380*	
Monochrome Television	NZ$	7	2	2	2	2	1	1	−13
	$M	**4**	**1**	**1**	**1**	**1**	**1**	**1**	
	No.th	*20*	*6*	*6*	*6*	*6*	*5*	*5*	
Video Recorders	NZ$	48	54	62	65	67	70	73	6
	$M	**28**	**29**	**33**	**35**	**36**	**38**	**39**	
	No.th	*108*	*112*	*124*	*130*	*135*	*141*	*147*	
Video Cameras & Camera Combinations	NZ$	32	35	33	36	38	40	42	4
	$M	**18**	**19**	**18**	**19**	**20**	**22**	**23**	
	No.th	*28*	*32*	*28*	*31*	*33*	*35*	*37*	
Total Video Equipment	NZ$	228	250	259	272	283	294	306	4
	$M	**132**	**134**	**139**	**146**	**152**	**158**	**165**	
Audio Equipment									
Portable Radios & Radio Recorders	NZ$	29	22	24	25	25	26	26	3
	$M	**17**	**12**	**13**	**13**	**13**	**14**	**14**	
	No.th	*503*	*485*	*544*	*560*	*575*	*590*	*610*	
Mains Radios & Combinations	NZ$	36	55	53	55	57	58	60	2
	$M	**21**	**30**	**28**	**30**	**31**	**31**	**32**	
	No.th	*268*	*281*	*304*	*320*	*335*	*350*	*365*	
Car Radios & Combinations	NZ$	11	17	19	20	21	21	22	5
	$M	**6**	**9**	**10**	**11**	**11**	**11**	**12**	
	No.th	*106*	*120*	*136*	*140*	*143*	*146*	*150*	
Tape Recorders	NZ$	15	17	15	15	16	17	17	0
	$M	**9**	**9**	**8**	**8**	**9**	**9**	**9**	
	No.th	*151*	*170*	*165*	*170*	*175*	*180*	*185*	
Record Players	NZ$	4	5	2	2	2	2	2	−17
	$M	**2**	**3**	**1**	**1**	**1**	**1**	**1**	
	No.th	*36*	*40*	*23*	*22*	*21*	*20*	*19*	
Compact Disc Players	NZ$	12	18	22	23	24	25	26	8
	$M	**7**	**10**	**12**	**12**	**13**	**13**	**14**	
	No.th	*50*	*68*	*79*	*85*	*90*	*95*	*100*	
Total Audio Equipment	NZ$	107	134	135	140	145	149	153	3
	$M	**62**	**72**	**73**	**75**	**78**	**80**	**82**	
Personal Consumer Equipment									
Electronic Musical Instruments	NZ$	9	10	10	11	11	11	11	2
	$M	**5**	**5**	**5**	**6**	**6**	**6**	**6**	
	No.th	*34*	*30*	*29*	*30*	*31*	*32*	*33*	
Electric/Electronic Watches	NZ$	21	18	23	24	25	26	27	8
	$M	**12**	**10**	**12**	**13**	**13**	**14**	**15**	
	No.th	*1145*	*1107*	*1426*	*1480*	*1550*	*1620*	*1690*	
Electric/Electronic Clocks	NZ$	6	5	6	6	6	7	7	7
	$M	**3**	**3**	**3**	**3**	**3**	**4**	**4**	
	No.th	*561*	*407*	*444*	*460*	*480*	*500*	*520*	
Electronic Flashlights	NZ$	1	1	1	1	1	1	1	0
	$M	**1**	**1**	**1**	**1**	**1**	**1**	**1**	
	No.th	*3*	*4*	*6*	*7*	*7*	*8*	*8*	
Total Personal Consumer Equipment	NZ$	37	34	40	42	43	45	46	6
	$M	**21**	**18**	**22**	**23**	**23**	**24**	**25**	
Total Consumer Market	NZ$	372	418	434	454	471	488	505	4
	$M	**215**	**225**	**233**	**244**	**253**	**262**	**272**	

NEW ZEALAND – MARKETS (cont)

COMPONENTS		1991	1992	1993	1994	1995	1996	1997	CAAGR %
Active Components									
Colour TV Tubes	NZ$	1	1	1	1	1	1	1	0
	$M	**1**	**1**	**1**	**1**	**1**	**1**	**1**	
Other Valves & Tubes	NZ$	2	2	1	2	2	2	2	0
	$M	**1**	**1**	**1**	**1**	**1**	**1**	**1**	
Diodes	NZ$	3	4	5	6	6	7	7	12
	$M	**2**	**2**	**3**	**3**	**3**	**4**	**4**	
Transistors	NZ$	6	11	10	11	12	13	14	5
	$M	**3**	**6**	**5**	**6**	**6**	**7**	**8**	
Other Discrete Semiconductors	NZ$	6	10	10	11	12	13	14	7
	$M	**3**	**5**	**5**	**6**	**6**	**7**	**8**	
ICs & Other Microcircuits	NZ$	18	26	47	58	69	78	88	28
	$M	**10**	**14**	**25**	**31**	**37**	**42**	**47**	
Total Active Components	NZ$	36	54	74	89	102	114	126	18
	$M	21	29	40	48	55	61	68	
Passive Components									
Electrolytic Capacitors	NZ$	1	2	3	3	4	4	5	20
	$M	**1**	**1**	**2**	**2**	**2**	**2**	**3**	
Other Fixed Capacitors	NZ$	5	9	8	9	10	10	11	4
	$M	**3**	**5**	**4**	**5**	**5**	**5**	**6**	
Variable Capacitors	NZ$	1	1	1	1	1	1	1	0
	$M	**1**	**1**	**1**	**1**	**1**	**1**	**1**	
Fixed Resistors	NZ$	2	2	2	2	2	2	2	0
	$M	**1**	**1**	**1**	**1**	**1**	**1**	**1**	
Variable Resistors	NZ$	2	3	4	4	5	5	5	11
	$M	**1**	**2**	**2**	**2**	**3**	**3**	**3**	
Other Connection Devices	NZ$	11	15	18	20	21	22	23	9
	$M	**6**	**8**	**10**	**11**	**11**	**12**	**12**	
Relays	NZ$	2	2	3	3	3	4	4	15
	$M	**1**	**1**	**2**	**2**	**2**	**2**	**2**	
Switches	NZ$	5	6	6	7	7	8	8	6
	$M	**3**	**3**	**3**	**4**	**4**	**4**	**4**	
Inductors	NZ$	6	8	9	10	10	11	12	8
	$M	**3**	**4**	**5**	**5**	**5**	**6**	**6**	
Printed Circuit Boards	NZ$	15	16	17	19	21	22	23	8
	$M	**9**	**9**	**9**	**10**	**11**	**12**	**12**	
Total Passive Components	NZ$	50	64	71	78	84	89	94	8
	$M	**29**	**34**	**38**	**42**	**45**	**48**	**51**	
Other Components									
Microphones	NZ$	4	3	3	3	3	3	3	0
	$M	2	2	2	2	2	2	2	
Loudspeakers	NZ$	13	14	14	15	15	16	16	3
	$M	**8**	**8**	**8**	**8**	**8**	**9**	**9**	
Amplifiers	NZ$	7	6	6	6	6	6	6	0
	$M	**4**	**3**	**3**	**3**	**3**	**3**	**3**	
Aerials	NZ$	3	4	5	6	7	8	9	18
	$M	2	2	**3**	**3**	**4**	**4**	**5**	

NEW ZEALAND – MARKETS (cont)

Other Components (cont)		**1991**	**1992**	**1993**	**1994**	**1995**	**1996**	**1997**	
Headphones & Earphones	NZ$	3	4	5	5	5	5	5	5
	$M	**2**	**2**	**3**	**3**	**3**	**3**	**3**	
Unrecorded Media	NZ$	41	40	44	47	49	50	51	5
	$M	**24**	**22**	**24**	**25**	**26**	**27**	**27**	
Accessories & Parts for Consumer Equipment	NZ$	16	16	17	18	19	19	20	5
	$M	**9**	**9**	**9**	**10**	**10**	**10**	**11**	
Total Other Components	NZ$	87	87	94	100	104	107	110	5
	$M	**50**	**47**	**51**	**54**	**56**	**58**	**59**	
Total Components Market	NZ$	173	205	239	267	290	310	330	10
	$M	**100**	**110**	**128**	**144**	**156**	**167**	**177**	

		1991	**1992**	**1993**	**1994**	**1995**	**1996**	**1997**	CAAGR
TOTAL MARKET–NEW ZEALAND	NZ$	1799	2149	2299	2450	2584	2724	2864	6
	$M	**1040**	**1155**	**1236**	**1317**	**1389**	**1465**	**1540**	

Notes: 1991 and 1992 are current figures at current exchange rates
1993 to 1997 are forecasts at 1992 constant values & exchange rates (ie. inflation is not included)
CAAGR column indicates true annual average growth between 1992 & 1997

4.6 PORTUGAL

BASIC STATISTICS

Land Area:		92 100 sq km
Population:		9.85 million (1992)
Civilian Employment:		4.7 million (1990)
of which:	- Agriculture	18%
	- Industry	34%
	- Other	48%
GNP per Capita:		US$8550 (1992)
TV Sets in Use:		1.7 million
Radio Sets in Use:		2.2 million
Cars per 1000 Inhabitants:		205 (1992)
Main Telephone Lines per 1000 Inhabitants:		255 (1992)

LEADING ECONOMIC INDICATORS (% increases)

	Actual				Forecast		
	1988	**1989**	**1990**	**1991**	**1992**	**1993**	**1994**
Growth of GDP (Constant Market Prices)	4.0	5.5	4.4	2.1	1.1	-2.0	1.6
Consumer Price Index	9.7	12.6	13.4	11.4	8.9	6.5	5.7
Industrial Production	6.1	5.2	8.6	0.0	-2.1	-4.0	1.5

Overview

Portugal lies at the south western tip of the continent of Europe on the west coast of the Iberian Peninsula. The major cities are Lisbon, the capital, and Oporto.

Portugal remains essentially an agricultural country, with tourism and wine playing important roles. The country's industrial base is narrow and the main manufacturing areas are textiles, footwear, furniture, ceramics/glassware, paper, cork, fertilisers and chemicals.

Portugal has been a member of the European Community (EC) since 1986. EC membership has already brought large net transfers from other EC countries and has dramatically boosted foreign investment inflows. Both are helping to finance more rapid modernisation of the economy.

GDP Growth

Following EC accession in 1986, Portugal became one of Europe's fastest growing economies with expansion averaging 4.6% a year, compared with an EC average of 3.3%. Growth has since slowed sharply, however, with the National Statistics Institute estimating that GDP fell by 2% or more in 1993. Growth of all components of domestic demand has slowed, and the real foreign balance has failed to benefit from currency devaluation.

The fall in real GDP, which began in the final quarter of 1992, accelerated in the first half of 1993 with a sharp drop in gross fixed investment in the first half of 1993. Private consumption growth, which underpinned output growth in 1992, has weakened sharply.

Led by a revival in exports and a related strengthening of fixed investment, activity should pick up in 1994 and gradually gather speed in 1995 as domestic demand recovers and European export markets strengthen. Lower interest rates and improved profitability should reinforce the projected re-acceleration of investment growth. Stronger growth of export markets should reduce the negative contribution to growth from the foreign balance.

Government Policy
The centre right Social Democratic Party, led by Prime Minister Anibal Cavaco Silva, has now been in power for over eight years. It first gained power and formed a minority government in November 1985. In the general election of July 1987, it increased its share of the vote and became the first of 17 administrations since the revolution to complete a full four year term of office, and maintained its share of the vote in the general election of October 1991.

Prime Minister Silva, who has led the government since 1985, has faced industrial unrest over stringent public sector pay ceilings, and his party was heavily defeated in the local elections of December 1993. He can point to substantial real wage gains since 1985, however, and will not have to relax economic policy for political reasons until 1995.

The 1994 budget aims to cut the state sector deficit to 6% of GDP. The measures include limiting of public spending, withdrawal of certain tax benefits and a tightening up of tax collections.

Prices & Inflation
The inflation rate fell from a peak of 33% in mid-1984 to 8% in early 1988 but has turned up sharply since then, reaching 13.4% in 1990. This was partly the result of factors beyond the government's control; initially, higher food prices due to unfavourable weather, and then higher oil prices due to the Gulf crisis.

Since 1990 inflation has fallen and the rapid rise in unemployment has helped to moderate nominal wage growth, reducing inflation further. Annual nominal wage growth in industry is estimated to have fallen to 6% in the first six months of 1993 from 11.7% the year before. This has permitted the annual increase in consumer prices to ease further to 6.5% in 1993, despite substantial currency depreciation.

The government aims to cut the average annual inflation rate to between 4% and 5.5% in 1994. Assuming only gradual progress here, this target implies an end 1994 rate of about 4.5% at most, which looks ambitious. Given the substantial slack in the goods market and the traditionally strong response of nominal wage growth to rising unemployment, consumer price inflation could nevertheless fall below 5% in 1995.

Industrial Outlook
Industrial production expanded strongly from 1988 to 1990 but output was flat in 1991 and declined from 1991 to 1993.

However, with the Ford and Volkswagen motor companies agreeing to build a massive Esc454 billion car factory in Portugal, foreign investors have clearly remained impressed by this location. Official statistics show a recent falling off in direct foreign investment in Portugal to Esc215 billion (US$14 billion) in the first nine months of 1993 compared with Esc423 billion (US$3.2 billion) in the corresponding period of 1992.

A threat might eventually come from Eastern Europe, where labour costs are also very low and there is plenty of surplus labour. This region currently has enormous problems of its own, however, and foreign direct investment in Portugal has so far held up as well as can be expected in a global recession.

A return to growth is forecast for 1994 accelerating in 1995 to reach around 4% in 1996.

Labour Market

Strong economic growth cut the unemployment rate from a peak of nearly 9% in 1985 to just over 4% in 1991-92. Since then, unemployment has risen and is expected to reach 6.5% in 1994.

The rapid rise in unemployment has helped to moderate nominal wage growth. Annual wage growth in industry is estimated to have fallen to 6% in the first six months of 1993 from 11.7% the year before. Unit labour costs are decelerating.

Wage pact talks have broken down, with the government offering pay rises of only 4% in 1994 but the unions looking for 5% plus the 1% implicit in the budget's 8% rise in tax thresholds. Wage restraint will now depend on the combination of the government's comfortable parliamentary majority and the weakness of the real economy.

Trade & Balance of Payments

Portuguese export volume growth has outperformed the EC average in every year but one (1991) since 1986. It has nevertheless lagged well behind import volume growth. The faster growth of import volume has reflected both the comparative buoyancy of domestic demand and a substantial erosion of international competitiveness.

The deterioration in the real trade balance has led to a sharp expansion in the nominal trade deficit since 1986 to US$9.4 billion in 1993. However, the balance of payments is not presenting a major problem at the moment, because EC transfers and other invisible income are containing the overall deficit on current account, and the deficit is being readily financed by foreign direct investment.

The current account deficit of US$600 million in 1993 is forecast to reduce during 1994 and 1995 until a small surplus of US$300 million is achieved in 1996.

ELECTRONICS INDUSTRY

STRUCTURE

In 1987 there were 102 members of the National Association for the Electronic and Electronics Industry (1987)

Main Production Activities: Consumer Equipment, Computer Peripherals, Telecommunications, Active and Passive Components.

Major Companies:

Alcatel
Centrel AEP
EFAEC
Electromecânica Portuguesa Preh
Grundig
Magnetic Peripherals Inc
Máqunas Eléctricas
Philips
Roederstein Electronica
Samsung
Siemens
Standard Electrica
Texas Instruments

Electronics Industry Employment: 13 450 (1990)

Portugal produces all types of electronic products with the most important being telecommunications, which accounts for 18% of total output; EDP equipment, mainly computer peripherals, accounting for 22% of total output; active components with 10%; passive components also 10% and consumer equipment 23%.

The industry experienced a real growth of 25% in both 1989 and 1990, the fastest growing being EDP equipment and control and instrumentation with a growth of over 80% in each of the two years. However, the electronics industry went into recession in 1991 which continued through 1992 and 1993. Total reduction was around 12% after allowing for inflation. The components sector was the worst affected with a real decline of 21%. A return to growth is forecast for 1994 with a real increase of 4.7%.

The Portuguese electronics industry is centred around the Lisbon, Operto, Braga and Setubal areas. The industry is highly dependent on foreign sources for the supply of raw materials and basic technology. A large proportion of the industry is dominated by multinationals, attracted to the region by relatively low labour costs and easy access to EC markets since 1986.

The main foreign-owned electronics companies operating in Portugal are Texas Instruments, Grundig, Alcatel, Siemens and Philips. Portuguese owned companies include Centrel AEP, EFAEC-Empresa Fabril de Máqunas Eléctricas and Electromecânica Portuguesa Preh.

There is strong infrastructure support for the electronics industry throughout Portugal. The Portuguese government is actively promoting investment, research and development, and staff training in the electronics sector.

ANIMEE, the National Association of Electrical and Electronic Equipment Industrialists, which represents over one hundred privately-owned companies, has established co-operative links with private and state-owned bodies, particularly universities. ANIMEE has also established CINEL, a training centre for the electronics industry, providing technical education for the strengthening, development and up-dating of the industry's work-force.

The government has recently launched an Information Technology and Electronics Development Programme (PITIE), to promote domestic and foreign investment in the electronics industry.

TRADE WITH THE EUROPEAN COMMUNITY

Values in current US$ millions

	Imports to the EC from Portugal			Exports from the EC To Portugal		
Product Group	**1990**	**1991**	**1992**	**1990**	**1991**	**1992**
EDP	34	32	32	301	371	402
Office Equipment	2	1	1	37	42	39
Control & Instruments	12	16	19	102	106	138
Medical	3	0	1	14	19	24
Industrial	1	2	2	18	21	31
Communications	18	16	22	41	39	73
Telecommunications	23	41	55	141	155	200
Consumer-Video	63	69	49	169	195	206
Consumer-Audio	166	186	272	61	74	89
Consumer-Personal	1	1	1	20	17	19
Active Components	167	134	97	41	73	80
Passive Components	139	158	146	74	155	170
Other Components	27	39	36	30	190	99
TOTAL	656	695	733	1049	1475	1570

Quantity in thousands of units

	Imports to the EC from Portugal			Exports from the EC to Portugal		
Product Group	**1990**	**1991**	**1992**	**1990**	**1991**	**1992**
Colour Television	241	259	169	480	653	439
Monochrome Television	53	21	26	19	11	63
Video Cassette Recorders	1	3	8	40	149	170

LIST OF SOURCES

Financial Times
Bank Economic Country Reports
Country Report & Country Profile — EIU
OECD Economic Survey
OECD Economic Outlook
Euromonitor
International Telecom Statistics — Siemens
World Automotive Statistics — SMMT
Predicasts F&S Index International
Electronic Times
Electronic News
Electronic Business
Electronic Engineering Times
Associaco Nacional dos industriais de material electrico e electrónico (ANIMEE)
Instituto do Investimento Estrangeiro/Foreign Investment Institute
Annual Statistics of Portugal

PORTUGAL – PRODUCTION

Table No.4.6.1

EDP		1991	1992	1993	1994	CAAGR %
Total EDP Production	EscM	49000	51000	48500	51400	0
	$M	**339**	**378**	**359**	**381**	

OFFICE EQUIPMENT		1991	1992	1993	1994	CAAGR %
Electronic Typewriters	EscM	800	700	600	550	-11
	$M	**6**	**5**	**4**	**4**	
	No.th	*20*	*20*	*18*	*17*	
Electronic Calculators, Cash Registers & Accounting Machines	EscM	500	400	350	330	-9
	$M	**3**	**3**	**3**	**2**	
	No.th	*60*	*50*	*45*	*43*	
Photocopiers	EscM	1100	1100	1140	1250	7
	$M	**8**	**8**	**8**	**9**	
	No.th	*9*	*9*	*10*	*11*	
Total Office Equipment Production	EscM	2400	2200	2090	2130	-2
	$M	**17**	**16**	**15**	**16**	

CONTROL & INSTRUMENTATION		1991	1992	1993	1994	CAAGR %
Total Control & Instrumentation	EscM	13000	14000	13700	14100	0
	$M	**90**	**104**	**101**	**104**	

MEDICAL & INDUSTRIAL		1991	1992	1993	1994	CAAGR %
X–Ray & Medical Equipment						
X–Ray & Radiation Equipment	EscM	210	320	350	380	9
	$M	**1**	**2**	**3**	**3**	
Other Electromedical Equipment	EscM	2100	2400	2500	2700	6
	$M	**15**	**18**	**19**	**20**	
Total X–Ray & Medical Equipment	EscM	2310	2720	2850	3080	6
	$M	**16**	**20**	**21**	**23**	

Industrial Equipment		1991	1992	1993	1994	CAAGR %
Signalling Equipment	EscM	700	600	620	660	5
	$M	**5**	**4**	**5**	**5**	
Induction & Dielectric Equipment	EscM	400	500	500	510	1
	$M	**3**	**4**	**4**	**4**	
Total Industrial Equipment	EscM	1100	1100	1120	1170	3
	$M	**8**	**8**	**8**	**9**	
Total Medical & Industrial Production	EscM	3410	3820	3970	4250	5
	$M	**24**	**28**	**29**	**31**	

COMMUNICATIONS & RADAR		1991	1992	1993	1994	CAAGR %
Total Communications & Radar	EscM	17000	16000	15700	16600	2
	$M	**118**	**119**	**116**	**123**	

PORTUGAL – PRODUCTION (cont)

TELECOMMUNICATIONS		1991	1992	1993	1994	CAAGR %
Total Telecommunications Production	EscM	42000	41000	36900	37600	−4
	$M	**291**	**304**	**273**	**279**	

CONSUMER		1991	1992	1993	1994	
Video Equipment						
Colour Television	EscM	16000	10000	10200	10700	3
	$M	**111**	**74**	**76**	**79**	
	No.th	*440*	*300*	*310*	*330*	
Monochrome Television	EscM	1100	1200	1250	1300	4
	$M	**8**	**9**	**9**	**10**	
	No.th	*100*	*110*	*120*	*130*	
Total Video Equipment	EscM	17100	11200	11450	12000	4
	$M	**118**	**83**	**85**	**89**	
Audio Equipment						
Portable Radios & Radio Recorders	EscM	1200	1100	1000	900	−10
	$M	**8**	**8**	**7**	**7**	
	No.th	*60*	*60*	*55*	*50*	
Car Radios	EscM	15600	31000	32000	33000	3
	$M	**108**	**230**	**237**	**244**	
	No.th	*920*	*1430*	*1470*	*1500*	
Main Radios	EscM	2420	2800	3000	3300	9
	$M	**17**	**21**	**22**	**24**	
	No.th	*180*	*194*	*205*	*220*	
Tape Recorders & Decks	EscM	5000	3500	3600	3800	4
	$M	**35**	**26**	**27**	**28**	
	No.th	*500*	*500*	*520*	*550*	
CD Players	PtsM	2700	2800	3000	3400	10
	$M	**19**	**21**	**22**	**25**	
	No.th	*212*	*330*	*350*	*400*	
Total Audio Equipment	EscM	26920	41200	42600	44400	4
	$M	**186**	**305**	**316**	**329**	
Personal Consumer Equipment						
Electric/Electronic Watches	PtsM	1000	900	880	920	1
	$M	**7**	**7**	**7**	**7**	
	No.th	*400*	*400*	*390*	*400*	
Electric/Electronic Clocks	EscM	600	400	390	410	1
	$M	**4**	**3**	**3**	**3**	
	No.th	*400*	*300*	*290*	*310*	
Personal Consumer Equipment Total	EscM	1600	1300	1270	1330	1
	$M	**11**	**10**	**9**	**10**	
Total Consumer Production	EscM	45620	53700	55320	57730	4
	$M	**316**	**398**	**410**	**428**	

PORTUGAL – PRODUCTION (cont)

COMPONENTS		1991	1992	1993	1994	CAAGR %
Active Components						
Diodes	EscM	350	500	510	550	5
	$M	**2**	**4**	**4**	**4**	
Transistors	EscM	550	500	520	550	5
	$M	**4**	**4**	**4**	**4**	
ICs & Other Microcircuits	EscM	21600	21000	21400	22700	4
	$M	**149**	**156**	**159**	**168**	
Total Active Components	EscM	22500	22000	22430	23800	4
	$M	**156**	**163**	**166**	**176**	
Passive Components						
Electrolytic Capacitors	EscM	890	850	830	860	1
	$M	**6**	**6**	**6**	**6**	
Other Fixed Capacitors	EscM	6600	6200	6000	6100	−1
	$M	**46**	**46**	**44**	**45**	
Fixed Resistors	EscM	1930	1750	1700	1750	0
	$M	**13**	**13**	**13**	**13**	
Variable Resistors	EscM	1700	1400	1350	1400	0
	$M	**12**	**10**	**10**	**10**	
Connectors	EscM	900	850	830	860	1
	$M	**6**	**6**	**6**	**6**	
Small Transformers & Inductors	EscM	5820	5700	5500	5600	−1
	$M	**40**	**42**	**41**	**41**	
Relays	EscM	5400	5500	5300	5400	−1
	$M	**37**	**41**	**39**	**40**	
Switches	EscM	780	750	700	720	−2
	$M	**5**	**6**	**5**	**5**	
Printed Circuit Boards	EscM	300	500	600	800	26
	$M	**2**	**4**	**4**	**6**	
Total Passive Components	EscM	24320	23500	22810	23490	0
	$M	**168**	**174**	**169**	**174**	
Other Components						
Loudspeakers	EscM	1700	1300	1250	1300	0
	$M	**12**	**10**	**9**	**10**	
Amplifiers	EscM	1300	1200	1150	1200	0
	$M	**9**	**9**	**9**	**9**	
Accessories & Parts for Consumer Equipment	EscM	1400	1200	1100	1100	−4
	$M	**10**	**9**	**8**	**8**	
Total Other Components	EscM	4400	3700	3500	3600	−1
	$M	**30**	**27**	**26**	**27**	
Total Components Production	EscM	51220	49200	48740	50890	2
	$M	**354**	**364**	**361**	**377**	

PORTUGAL – PRODUCTION (cont)

		1991	1992	1993	1994	CAAGR %
TOTAL PRODUCTION–PORTUGAL	EscM	223650	230920	224920	234700	1
	$M	**1548**	**1711**	**1666**	**1739**	

Notes: 1991 and 1992 are current figures at current exchange rates
1993 & 1994 are forecasts at 1992 constant values & exchange rates (ie inflation is not included)
CAAGR column indicates true annual average growth between 1992 & 1994

Portugal Trade Balance 1992

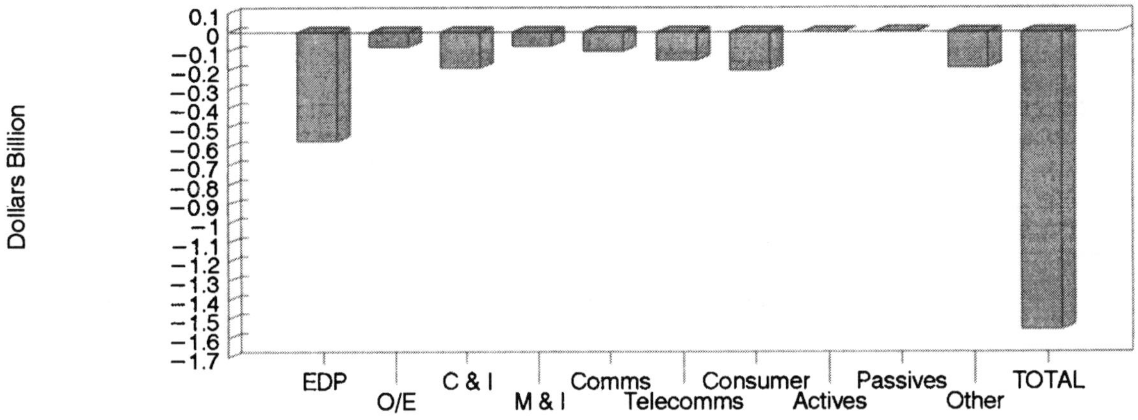

PORTUGAL – MARKETS

Table No.4.6.2

EDP		1991	1992	1993	1994	1995	1996	1997	CAAGR %
Total EDP Market	EscM	128837	128732	121000	125800	136000	150000	165000	5
	$M	**892**	**954**	**896**	**932**	**1007**	**1111**	**1222**	

OFFICE EQUIPMENT		1991	1992	1993	1994	1995	1996	1997	CAAGR %
Electronic Typewriters	EscM	1757	1497	1340	1300	1270	1240	1200	–4
	$M	**12**	**11**	**10**	**10**	**9**	**9**	**9**	
	No.th	*52*	*47*	*45*	*46*	*47*	*48*	*49*	
Electronic Calculators, Cash Registers & Accounting Machines	EscM	2921	3092	2900	2900	3050	3200	3400	2
	$M	**20**	**23**	**21**	**21**	**23**	**24**	**25**	
	No.th	*1390*	*1690*	*1650*	*1680*	*1800*	*1900*	*2000*	
Photocopiers	EscM	7797	8026	7500	7650	8000	8500	9000	2
	$M	**54**	**59**	**56**	**57**	**59**	**63**	**67**	
	No.th	*34*	*35*	*34*	*36*	*38*	*41*	*44*	
Total Office Equipment Market	EscM	12475	12615	11740	11850	12320	12940	13600	2
	$M	**86**	**93**	**87**	**88**	**91**	**96**	**101**	

CONTROL & INSTRUMENTATION		1991	1992	1993	1994	1995	1996	1997	CAAGR %
Total Control & Instrumentation	EscM	36944	39455	37900	38600	41000	44200	48000	4
	$M	**256**	**292**	**281**	**286**	**304**	**327**	**356**	**0**

MEDICAL & INDUSTRIAL		1991	1992	1993	1994	1995	1996	1997	CAAGR %
X–Ray & Medical Equipment									
X–Ray & Radiation Equipment	EscM	2042	1907	1800	1850	2000	2200	2400	5
	$M	**14**	**14**	**13**	**14**	**15**	**16**	**18**	
Other Electromedical Equipment	EscM	4823	4869	4640	4840	5360	6060	6860	7
	$M	**33**	**36**	**34**	**36**	**40**	**45**	**51**	
Hearing Aids	EscM	195	212	200	210	230	250	270	5
	$M	**1**	**2**	**1**	**2**	**2**	**2**	**2**	
Total X–Ray & Medical Equipment	EscM	7060	6988	6640	6900	7590	8510	9530	6
	$M	**49**	**52**	**49**	**51**	**56**	**63**	**71**	
Industrial Equipment									
Signalling Equipment	EscM	5009	6292	5980	6100	6500	6900	7300	3
	$M	**35**	**47**	**44**	**45**	**48**	**51**	**54**	
Induction & Dielectric Equipment	EscM	884	743	700	710	730	760	800	1
	$M	**6**	**6**	**5**	**5**	**5**	**6**	**6**	
Total Industrial Equipment	EscM	5893	7035	6680	6810	7230	7660	8100	3
	$M	**41**	**52**	**49**	**50**	**54**	**57**	**60**	
Total Medical & Industrial Market	EscM	12953	14023	13320	13710	14820	16170	17630	5
	$M	**90**	**104**	**99**	**102**	**110**	**120**	**131**	

COMMUNICATIONS & RADAR		1991	1992	1993	1994	1995	1996	1997	CAAGR %
Total Communications & Radar	EscM	25417	29565	29000	29800	31600	34200	37000	5
	$M	**176**	**219**	**215**	**221**	**234**	**253**	**274**	

PORTUGAL – MARKETS (cont)

TELECOMMUNICATIONS		1991	1992	1993	1994	1995	1996	1997	CAAGR %
Total Telecommunications Market	EscM	56718	61076	58600	58000	59200	61000	62800	1
	$M	**393**	**452**	**434**	**430**	**439**	**452**	**465**	

CONSUMER		1991	1992	1993	1994	1995	1996	1997	CAAGR %
Video Equipment									
Colour Television	EscM	33843	27286	27080	27800	29200	31080	33200	4
	$M	**234**	**202**	**201**	**206**	**216**	**230**	**246**	
	No.th	*849*	*731*	*725*	*740*	*770*	*820*	*870*	
Monochrome Television	EscM	1350	1124	1050	1000	950	900	900	-4
	$M	**9**	**8**	**8**	**7**	**7**	**7**	**7**	
	No.th	*92*	*81*	*75*	*71*	*67*	*64*	*62*	
Video Recorders	EscM	13199	12565	12000	12400	13000	13800	14500	3
	$M	**91**	**93**	**89**	**92**	**96**	**102**	**107**	
	No.th	*290*	*304*	*290*	*300*	*310*	*330*	*350*	
Video Cameras & Camera Combinations	EscM	6206	8236	8100	8500	9000	9500	10000	4
	$M	**43**	**61**	**60**	**63**	**67**	**70**	**74**	
	No.th	*66*	*88*	*87*	*92*	*98*	*104*	*110*	
Total Video Equipment	EscM	54598	49211	48230	49700	52150	55280	58600	4
	$M	**378**	**365**	**357**	**368**	**386**	**409**	**434**	
Audio Equipment									
Portable Radios & Radio Recorders	EscM	3662	2853	2780	2830	2900	3000	3100	2
	$M	**25**	**21**	**21**	**21**	**21**	**22**	**23**	
	No.th	*1396*	*1198*	*1150*	*1160*	*1180*	*1200*	*1240*	
Mains Radios & Combinations (inc Clock Radios)	EscM	6695	7363	7200	7300	7600	8000	8400	3
	$M	**46**	**55**	**53**	**54**	**56**	**59**	**62**	
	No.th	*789*	*777*	*750*	*760*	*780*	*820*	*850*	
Car Radios & Combinations	EscM	1271	1216	1160	1150	1200	1300	1450	4
	$M	**9**	**9**	**9**	**9**	**9**	**10**	**11**	
	No.th	*154*	*149*	*140*	*137*	*140*	*150*	*160*	
Tape Recorders & Decks	EscM	4794	4851	4700	4800	4950	5200	5450	2
	$M	**33**	**36**	**35**	**36**	**37**	**39**	**40**	
	No.th	*1275*	*1129*	*1100*	*1150*	*1200*	*1300*	*1400*	
Record Players & Decks	EscM	496	431	410	390	380	370	360	-4
	$M	**3**	**3**	**3**	**3**	**3**	**3**	**3**	
	No.th	*62*	*50*	*49*	*48*	*47*	*46*	*45*	
Compact Disc Players	EscM	2989	3233	3300	3470	3700	3900	4100	5
	$M	**21**	**24**	**24**	**26**	**27**	**29**	**30**	
	No.th	*220*	*194*	*200*	*210*	*220*	*230*	*240*	
Total Audio Equipment	EscM	19907	19947	19550	19940	20730	21770	22860	3
	$M	**138**	**148**	**145**	**148**	**154**	**161**	**169**	
Personal Consumer Equipment									
Electronic Musical Instruments	EscM	1506	1271	1200	1220	1250	1300	1350	1
	$M	**10**	**9**	**9**	**9**	**9**	**10**	**10**	
	No.th	*79*	*74*	*73*	*75*	*77*	*80*	*83*	
Electric/Electronic Watches	EscM	8269	9012	8770	8950	9330	9800	10300	3
	$M	**57**	**67**	**65**	**66**	**69**	**73**	**76**	
	No.th	*5382*	*5853*	*5700*	*5800*	*6000*	*6200*	*6400*	
Electric/Electronic Clocks	EscM	1711	1436	1400	1430	1480	1550	1630	3
	$M	**12**	**11**	**10**	**11**	**11**	**11**	**12**	
	No.th	*1580*	*1581*	*1550*	*1600*	*1650*	*1700*	*1800*	

PORTUGAL – MARKETS (cont)

		1991	1992	1993	1994	1995	1996	1997	
Electronic Flashlights	EscM	171	150	145	148	154	160	165	2
	$M	**1**	**1**	**1**	**1**	**1**	**1**	**1**	
	No.th	*61*	*92*	*90*	*92*	*96*	*98*	*100*	
Total Personal Consumer Equipment	EscM	11657	11869	11515	11748	12214	12810	13445	3
	$M	**81**	**88**	**85**	**87**	**90**	**95**	**100**	
Total Consumer Market	EscM	86162	81027	79295	81388	85094	89860	94905	3
	$M	**596**	**600**	**587**	**603**	**630**	**666**	**703**	

COMPONENTS		**1991**	**1992**	**1993**	**1994**	**1995**	**1996**	**1997**	CAAGR %
Active Components									
Colour Television Tubes	EscM	3892	2018	2070	2170	2300	2400	2500	4
	$M	**27**	**15**	**15**	**16**	**17**	**18**	**19**	
Monochrome Television Tubes	EscM	127	121	126	131	136	140	145	4
	$M	**1**	**1**	**1**	**1**	**1**	**1**	**1**	
Other Valves & Tubes	EscM	1123	687	670	690	710	730	750	2
	$M	**8**	**5**	**5**	**5**	**5**	**5**	**6**	
Diodes	EscM	1394	1240	1200	1260	1320	1380	1440	3
	$M	**10**	**9**	**9**	**9**	**10**	**10**	**11**	
Transistors	EscM	1488	1445	1400	1450	1500	1550	1600	2
	$M	**10**	**11**	**10**	**11**	**11**	**11**	**12**	
Other Discrete Semiconductors	EscM	1530	1262	1260	1320	1380	1440	1500	4
	$M	**11**	**9**	**9**	**10**	**10**	**11**	**11**	
ICs & Other Microcircuits	EscM	15727	15482	15100	16100	17600	19300	21200	6
	$M	**109**	**115**	**112**	**119**	**130**	**143**	**157**	
Total Active Components	EscM	25281	22255	21826	23121	24946	26940	29135	6
	$M	**175**	**165**	**162**	**171**	**185**	**200**	**216**	
Passive Components									
Electrolytic Capacitors	EscM	969	925	880	920	970	1020	1070	3
	$M	**7**	**7**	**7**	**7**	**7**	**8**	**8**	
Other Fixed Capacitors	EscM	2829	3002	2900	3000	3150	3300	3450	3
	$M	**20**	**22**	**21**	**22**	**23**	**24**	**26**	
Variable Capacitors	EscM	207	106	100	104	108	112	116	2
	$M	**1**	**1**	**1**	**1**	**1**	**1**	**1**	
Fixed Resistors	EscM	871	870	830	850	880	910	940	2
	$M	**6**	**6**	**6**	**6**	**7**	**7**	**7**	
Variable Resistors	EscM	1698	1661	1600	1650	1700	1750	1800	2
	$M	**12**	**12**	**12**	**12**	**13**	**13**	**13**	
Connectors	EscM	9245	9389	9000	9400	9900	10400	11000	3
	$M	**64**	**70**	**67**	**70**	**73**	**77**	**81**	
Small Transformers & Inductors	EscM	1651	1567	1500	1540	1590	1640	1690	
	$M	**11**	**12**	**11**	**11**	**12**	**12**	**13**	
Relays	EscM	1465	1270	1200	1240	1280	1320	1360	1
	$M	**10**	**9**	**9**	**9**	**9**	**10**	**10**	
Switches	EscM	1080	1027	980	1010	1040	1070	1100	1
	$M	**7**	**8**	**7**	**7**	**8**	**8**	**8**	
Printed Circuit Boards	EscM	3569	3362	3260	3430	3700	4000	4300	5
	$M	**25**	**25**	**24**	**25**	**27**	**30**	**32**	
Total Passive Components	EscM	23584	23179	22250	23144	24318	25522	26826	3
	$M	**163**	**172**	**165**	**171**	**180**	**189**	**199**	

PORTUGAL – MARKETS (cont)

Other Components		**1991**	**1992**	**1993**	**1994**	**1995**	**1996**	**1997**	
Microphones	EscM	246	322	300	305	310	320	330	0
	$M	**2**	**2**	**2**	**2**	**2**	**2**	**2**	
Loudspeakers	EscM	3732	3657	3400	3500	3600	3700	3800	1
	$M	**26**	**27**	**25**	**26**	**27**	**27**	**28**	
Amplifiers	EscM	1794	1896	1850	1900	2000	2100	2200	3
	$M	**12**	**14**	**14**	**14**	**15**	**16**	**16**	
Aerials	EscM	2215	3788	3800	3900	4100	4300	4500	4
	$M	**15**	**28**	**28**	**29**	**30**	**32**	**33**	
Unrecorded Media	EscM	7234	7463	7000	7200	7500	7800	8000	1
	$M	**50**	**55**	**52**	**53**	**56**	**58**	**59**	
Accessories & Parts for Consumer Equipment	EscM	11413	12165	11200	11300	11400	11600	11800	-1
	$M	**79**	**90**	**83**	**84**	**84**	**86**	**87**	
Total Other Components	EscM	26634	29291	27550	28105	28910	29820	30630	1
	$M	**184**	**217**	**204**	**208**	**214**	**221**	**227**	
Total Components Market	EscM	75499	74725	71626	74370	78174	82282	86591	3
	$M	**522**	**554**	**531**	**551**	**579**	**609**	**641**	

		1991	**1992**	**1993**	**1994**	**1995**	**1996**	**1997**	CAAGR
TOTAL MARKET – PORTUGAL	EscM	435005	441218	422481	433518	458208	490652	525526	4
	$M	**3010**	**3268**	**3129**	**3211**	**3394**	**3634**	**3893**	

Notes: 1991 and 1992 are current figures at current exchange rates
1993 to 1997 are forecasts at 1992 constant values & exchange rates (ie. inflation is not included)
CAAGR column indicates true annual average growth between 1992 & 1997

4.7 PUERTO RICO

BASIC STATISTICS

Land Area:		8897 sq km
Population:		3.4 million (1992)
Civilian Employment:		1 013 000 (1993)
of which:	- Agriculture	3%
	- Industry	23%
	- Other	74%
GDP per Capita:		US$8895 (1990)
TV sets in Use:		830 000
Radio sets in Use:		2 million
Cars per 1000 Inhabitants:		397 (1992)

LEADING ECONOMIC INDICATORS (% increases)

	Actual				Forecast		
	1988	1989	1990	1991	1992	1993	1994
Growth of GDP (constant market prices)	4.9	3.7	2.2	1.3	1.0	3.0	3.0
Consumer Price Index	2.9	3.7	5.7	2.9	2.5	2.7	3.0

Overview

Puerto Rico, a Spanish colony until 1898, became a US protectorate following the Spanish-American war of that year. Since 1952 Puerto Rico has had commonwealth status with the US. With the formal establishment of the commonwealth, Puerto Rico acquired a large measure of self government, setting its own education policies, establishing its own internal taxation procedures and its own criminal code. However, economically the island is dependant upon the US for federal funds and private industrial capital. There is a large outflow of profits and interest payments to firms and creditors on the mainland. Puerto Rico is represented in Washington by a resident commissioner but its citizens have only recently been granted the right to vote in certain federal elections. There are US bases at Roosevelt Roads, Santiago, Sabana Seca and Vieques which have recently been modernised and strengthened; the Roosevelt naval base is particularly important being the US Navy's largest base worldwide.

The climate is sub-tropical, the language Spanish and English, the currency the US dollar.

GDP Growth

Puerto Rico's economic growth since the end of the 2nd world war has been exceptional. GDP growth rates from 1947 to 1973 averaged about 7% p.a. in real terms. However, there was a marked drop in the recession years of 1973-1977 when GDP averaged 1.7% p.a. The 1979/80 oil price increases and anti-inflation policies in the US pointed the Puerto Rico economy towards another period of negative growth, but after 2 years of recession the economy achieved positive growth from 1984 to 1989. After Hurricane Hugo caused severe damage in September 1989 the reconstruction led to a short-lived boom, but a slow-down in the mainland US economy had its effect on the island's manufacturing industries and the growth rate in 1990 was much lower at only 2.2%. The continued sluggishness of the US recovery has delayed any improvements and there has been a spate of bankruptcies. GDP in 1992 was only 1%. With the US economy now recovering from recession, in the short term the local economy should see steady growth following on from around 3% growth in the fiscal year to June 1993. In the medium term the question of statehood may again be raised with adverse effects on investment, as has been the case in the last three years. However, this debate should not resume for some time and so there should be no impact on the economy in 1994.

Government Policy

Since 1968 power has been shared between two main political parties with an elected government and a national legislation. The Puerto Popular Democratico (PPD) governor proposed a referendum on the state of the island. The options were status quo, statehood or independence. The PPD favoured the status quo while the opposition favoured statehood. If the later were chosen it was expected that after a transitional period, Puerto Rico individuals and firms would become liable for federal taxes and federal benefits. At present profits earned by US companies Puerto Rico operations are exempt from US income tax and is regarded as by far the largest investment incentive for the island. The plebiscites was held in 1993 and the decision by the electorate was to stay with the current constitution status and not to seek to become the 51st state in the US. This has, for the time being at least, removed major political uncertainty over the island's future.

Prices & Inflation

Inflation follows the trend in US prices. Salaries have more than kept pace with the cost of living. The federal minimum wage applies and, with so many workers earning little more than this level, any rise in the federal minimum is strongly felt. The island was partially exempt from the increase in minimum rates passed by the US Congress in 1990, although wages below that level will be adjusted towards the new minimum. Inflation in 1994 is forecast to rise slightly to 3% from 2.7% in 1993.

Industrial Outlook

In the span of 40 years the Puerto Rico economy has been transformed from a traditional agricultural one into an industrialised economy based on manufacturing, commerce, construction, tourism and government. Although the manufacturing sector as a whole has been a major growth area throughout the post-war years, inevitably there have been changes in the relative importance of various industry groups. Textiles and clothing have lost ground in the face of Far East competition. Since 1970 the chemical and machinery industries have shown the most rapid growth. Substantial petroleum and pharmaceutical industries have also been developed; Puerto Rico is now the worlds leading exporter of pharmaceutical products.

Labour Market

Population growth of the island has outstripped employment which has led to massive emigration to the US mainland. Unemployment climbed to a peak of 25% in 1983. Since then, however, it has dropped slightly with the creation of new jobs in manufacturing and government sectors. Unemployment is still at least twice that in the US mainland and in March 1993 was 17.2% against 16.8% in March 1992. Although unemployment is highest among the young and less well qualified, it is also high for the older age groups and college trained workers. The sectors most adversely affected have been construction and agriculture.

Trade & Balance of Payments
Puerto Rico continues to send most of its exports to the US and to obtain most of its imports from the mainland, but the island's trade surplus with the mainland has declined. In the July 1992 to March 1993 period, exports to the US were down by 9.4% on a year earlier while imports were up 5.2%, to leave the island's trade balance with the US at US$4271 million, down 26%. Exports to the US destinations rose by 2.9% but imports were up 3.7%. The result was that the trade supply for the island fell by 37% to US$2632 million.

ELECTRONICS INDUSTRY

STRUCTURE

Information on the size of the Puerto Rico electronics industry has been sparse but detailed statistics of trade between the country and the US now enable a realistic estimate to be made.

In 1992 output was US$3 billion of which 58% was production of computer equipment and computer parts.

The main problem in accessing the electronics industry is that Puerto Rico still uses the 1987 SIC coding system which effectively excluded any detail on the computer industry.

In 1993 there were 108 plants producing electronic or electrical equipment. This excludes plants engaged in production of computer equipment and scientific instruments including electronic instruments.

Of these, the number producing electronic products (excluding computers) were:-

Consumer	1
Telecommunications	6
Radio Communications	4
Industrial	8
Components	23
TOTAL	42

Employment in the electronic and electrical sector was 17.3 million, a reduction of 14% since 1990. In addition are employees in the computer sector for which no separate figures exist. The electronic and electrical sector accounts for 12% of the total of 149 000 workers in the manufacturing sector in Puerto Rico.

Some of the major electronic companies are:

Local Company	**Parent**
Banctec Puerto Rico Inc	Banctec Inc
Caribe General Electric Products Inc	General Electric Company
Challenger Caribbean Corp	Thomas & Betts Corp
Checkpoint Systems of P.R.	Checkpoint Systems Inc
Cutler Hammer de P.R. Inc	Eaton Corporation
Digital Communication Associates(P.R)	Digital Communications Assoc
DSC of Puerto Rico Inc	DSC Communications Corp

General Instruments (P.R.) Inc	General Instruments Corp
GTE Control Devices of P.R.Inc	GTE Corporation
Intel Puerto Rico Inc	Intel Corporation
Matsushita Electric of P.R. Inc	Matsushita Electric Corp
Microsoft Puerto Rico Inc	Microsoft Corporation
Milgray Electronic P.R. Inc	Milgray Electronics Inc
Molex Caribe Inc	Molex Inc
Motorola Telcarpo de P.R. Inc	Motorola Inc
Northern Telecom (Puerto Rico) Inc	Northern Telecom Ltd
Osram Sylvania Puerto Rico Corp	GTE Corporation
Perkin-Elmer Caribbean Corp	The Perkin-Elmer Corp
Power Parts Inc	R.E. Phelon Co Inc
Productos Circuitos de P.R. Div of W.	Westinghouse Electric Corp
Semiconductores Powerex Inc	Powerex Inc
Suttle Caribe Inc	Communications Systems Inc
Telecom Solutions de P.R. Inc	Silicon General Inc
Telular International Inc	Telular Inc
Timeplex Caribe Ltd	Unisys Corporation

The electronic products manufactured on the island include:

Computer Equipment	Transformers
CB Radio Scanners	Cassettes
Clock Radios	Circuit Breakers
Detection Equipment	Communications Equipment
Electronic Modules	Electronic Controls
Fire Alarms	Microwave Filters
Microphones	Photoelectric Cells
Pacemakers	Television Parts
Printed Circuit Boards	Voltage Regulators
Switches	

The electronics industry in Puerto Rico is largely controlled by US companies, with 75% of manufacturing facilities being subsidiaries of US firms. Sixteen of the US firms with plants operating in Puerto Rico are Fortune 500 companies.

The majority of trade is with the United States. In 1993 88% of exports of electronic and electrical products, excluding computer equipment, and 81% of imports were with the US. It is believed that if computer equipment were included the proportion would be much higher.

In compiling the estimates of markets and production on the following pages, US trade statistics have been used as the Puerto Rican statistics are not sufficiently detailed.

LIST OF SOURCES

Financial Times
Bank Economic Country Reports
Country Reports and Country Profile — EIU
International Telecom Statistics — Siemens
World Automotive Statistics — SMMT
Predicasts F & S Index International
Euromonitor
Economic Development Administration — Commonwealth of Puerto Rico
Shipment of Merchandise from to US to Puerto Rico — US Dept of Commerce
Shipment of Merchandise from Puerto Rico to the US — US Dept of Commerce

PUERTO RICO – PRODUCTION

Table No.4.7.1

EDP		1991	1992	1993	1994	CAAGR %
Complete Systems	$M	1200	830	800	800	-2
Accessories & Parts	$M	900	940	1020	1100	8
Total EDP Production	$M	2100	1770	1820	1900	4

OFFICE EQUIPMENT		1991	1992	1993	1994	CAAGR
Total Office Equipment Production	$M	0	0	0	0	

CONTROL & INSTRUMENTATION		1989	1990	1991	1992	CAAGR
Instrumentation Total	$M	203	190	194	198	2
Accessories & Parts	$M	12	10	10	10	0
Total Control & Instrumentation	$M	215	200	204	208	2

MEDICAL & INDUSTRIAL		1991	1992	1993	1994	CAAGR %
Total X-Ray & Medical Equipment	$M	100	136	150	165	10
Total Industrial Equipment	$M	38	40	42	44	% 5
Total Medical & Industrial Production	$M	138	176	192	209	9

COMMUNICATIONS & RADAR		1991	1992	1993	1994	CAAGR %
Radio Comms & Public Broadcast	$M	140	235	282	338	20
Accessories & Parts	$M	7	11	13	16	21
Total Communications & Radar	$M	147	246	295	354	20

TELECOMMUNICATIONS		1991	1992	1993	1994	CAAGR %
Telecommunications Equipment	$M	80	70	73	76	4
Accessories & Parts	$M	30	10	11	12	10
Total Telecommunications Production	$M	110	80	84	88	5

PUERTO RICO – PRODUCTION (cont)

CONSUMER		1991	1992	1993	1994	
Video Equipment						
Colour TV Receivers	$M	48	45	43	41	–5
	No.th	*160*	*210*	*200*	*190*	
Total Video Equipment	$M	48	45	43	41	
Audio Equipment						
Radios	$M	4	4	4	4	0
	No.th	*10*	*10*	*11*	*12*	
Total Audio Equipment	$M	4	4	4	4	0
Personal Consumer Equipment						
Electronic Watches (including movements)	$M	18	15	14	14	–3
	No.th	*140*	*150*	*160*	*170*	
Total Personal Consumer Equipment	$M	18	15	14	14	
Total Consumer Production	$M	70	64	61	59	–4

COMPONENTS		1991	1992	1993	1994	CAAGR %
Active Components						
Discretes	$M	20	12	13	14	8
ICs & Other Microcircuits	$M	100	70	77	85	10
Total Active Components	$M	120	82	90	99	10
Passive Components						
Capacitors	$M	30	25	26	27	4
Resistors	$M	10	7	7	6	–7
Connectors	$M	50	45	46	47	2
Relays	$M	15	12	12	13	4
Switches	$M	12	10	10	11	5
Inductors	$M	5	3	3	3	0
Printed Circuit Boards	$M	50	40	45	50	12
Total Passive Components	$M	172	142	149	157	5
Other Components						
Microphones, Loudspeakers & Amplifiers	$M	30	25	26	27	4
Aerials	$M	10	8	8	8	0
Unrecorded Media	$M	210	210	225	240	7
Accessories & Parts for Consumer Equipment	$M	50	48	47	46	–2
Total Other Components	$M	300	291	306	321	5
Total Components Production	$M	592	515	545	577	6

PUERTO RICO – PRODUCTION (cont)

		1991	1992	1993	1994	CAAGR %
TOTAL PRODUCTION–PUERTO RICO	**$M**	**3372**	**3051**	**3201**	**3395**	5

Notes: 1991 and 1992 are current figures
1993 & 1994 are forecasts at 1992 constant values (ie inflation is not included)
CAAGR column indicates true annual average growth between 1992 & 1994

Puerto Rico Trade Balance 1992

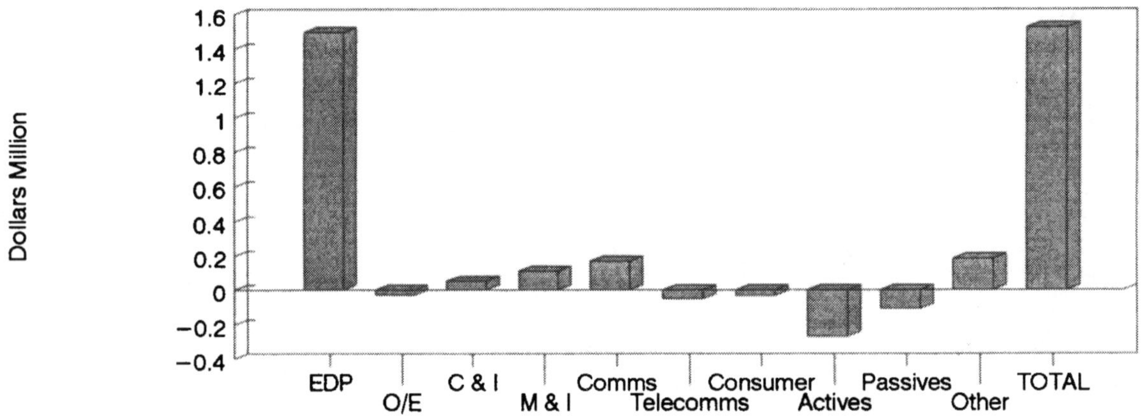

PUERTO RICO – MARKETS

Table No.4.7.2

EDP		1991	1992	1993	1994	1995	1996	1997	CAAGR %
Complete Systems	$M	217	226	235	245	255	260	265	
Accessories & Parts	$M	47	51	55	60	65	70	75	8
Total EDP Market	$M	264	277	290	305	320	330	340	4

OFFICE EQUIPMENT		1991	1992	1993	1994	1995	1996	1997	CAAGR %
Electronic Typewriters	$M	4	3	3	3	3	2	2	−8
	No.th	*15*	*13*	*14*	*14*	*15*	*15*	*15*	
Electronic Calculators	$M	2	2	2	2	2	2	2	0
	No.th	*300*	*350*	*360*	*370*	*380*	*385*	*390*	
Electronic Cash Registers	$M	1	1	1	1	1	1	1	0
	No.th	*3*	*3*	*3*	*4*	*4*	*4*	*4*	
Photocopiers	$M	17	21	22	22	23	24	24	3
	No.th	*10*	*16*	*17*	*18*	*19*	*20*	*21*	
Total Office Equipment Market	$M	24	27	28	28	29	29	29	1

CONTROL & INSTRUMENTATION		1991	1992	1993	1994	1995	1996	1997	CAAGR %
Instrumentation Total	$M	62	65	72	79	85	90	94	8
Accessories & Parts	$M	69	83	88	92	96	100	104	5
Total Industrial & Process Control	$M	131	148	160	171	181	190	198	6

MEDICAL & INDUSTRIAL		1991	1992	1993	1994	1995	1996	1997	CAAGR %
Total X–Ray & Medical Equip.	$M	42	46	51	55	60	65	70	9
Total Industrial Equipment	$M	18	19	20	21	22	23	24	5
Total Medical & Industrial Market	$M	60	65	71	76	82	88	94	8

COMMUNICATIONS & RADAR		1991	1992	1993	1994	1995	1996	1997	CAAGR %
Radar & Navigational Aids	$M	7	8	8	7	7	7	7	−3
Radio Comms & Public Broadcast	$M	53	59	66	74	82	89	95	10
Accessories & Parts	$M	11	10	11	12	13	14	15	8
Total Communications & Radar	$M	71	77	85	93	102	110	117	9

TELECOMMUNICATIONS		1991	1992	1993	1994	1995	1996	1997	CAAGR %
Telecommunications Equipment	$M	47	37	39	41	40	39	38	1
Accessories & Parts	$M	52	93	95	96	95	94	91	0
Total Telecommunications Market	$M	99	130	134	137	135	133	129	0

PUERTO RICO – MARKETS (cont)

CONSUMER		1991	1992	1993	1994	1995	1996	1997	CAAGR %
Video Equipment									
Colour TV Receivers	**$M**	**26**	**37**	**41**	**45**	**47**	**49**	**51**	7
	No.th	*142*	*150*	*170*	*190*	*200*	*210*	*220*	
Monochrome TV Receivers	**$M**	**5**	**5**	**4**	**4**	**4**	**4**	**3**	−10
	No.th	*40*	*48*	*47*	*46*	*45*	*44*	*43*	
VCR/Player	**$M**	**10**	**12**	**14**	**15**	**16**	**17**	**18**	8
	No.th	*30*	*41*	*48*	*51*	*54*	*57*	*61*	
Video Cameras	**$M**	**4**	**8**	**9**	**10**	**11**	**12**	**13**	10
	No.th	*6*	*13*	*15*	*17*	*18*	*20*	*22*	
Total Video Equipment	**$M**	**45**	**62**	**68**	**74**	**78**	**82**	**85**	
Audio Equipment									
Radios	**$M**	**8**	**10**	**11**	**12**	**12**	**13**	**14**	7
	No.th	*125*	*165*	*180*	*195*	*205*	*220*	*235*	
Record Players	**$M**	**2**	**1**	**1**	**1**	**1**	**0**	**0**	−100
	No.th	*6*	*2*	*2*	*2*	*2*	*1*	*1*	
Tape Recorders	**$M**	**5**	**5**	**5**	**5**	**6**	**6**	**6**	4
	No.th	*100*	*102*	*108*	*112*	*115*	*118*	*120*	
CD Players	**$M**	**9**	**7**	**8**	**9**	**9**	**10**	**10**	7
	No.th	*50*	*53*	*65*	*74*	*82*	*90*	*100*	
Total Audio Equipment	**$M**	**24**	**23**	**25**	**27**	**28**	**29**	**30**	5
Personal Consumer Equipment									
Electronic Musical Instruments	**$M**	**1**	**1**	**1**	**1**	**1**	**1**	**1**	0
	No.th	*3*	*4*	*4*	*5*	*5*	*5*	*5*	
Electronic Watches (including movements)	**$M**	**6**	**10**	**11**	**12**	**13**	**13**	**14**	7
	No.th	*501*	*576*	*633*	*690*	*730*	*770*	*810*	
Electronic Clocks (including movements)	**$M**	**1**	**1**	**1**	**1**	**1**	**1**	**1**	0
	No.th	*109*	*106*	*114*	*122*	*128*	*133*	*136*	
Total Personal Consumer Equip.	**$M**	**8**	**12**	**13**	**14**	**15**	**15**	**16**	6
Total Consumer Market	**$M**	**77**	**97**	**106**	**115**	**121**	**126**	**131**	6

COMPONENTS		1991	1992	1993	1994	1995	1996	1997	CAAGR %
Active Components									
Valves & Tubes	**$M**	**1**	**1**	**1**	**1**	**1**	**1**	**1**	
Discrete Semiconductors	**$M**	**41**	**43**	**46**	**49**	**51**	**53**	**55**	5
ICs & Other Microcircuits	**$M**	**276**	**307**	**339**	**375**	**411**	**446**	**479**	
Total Active Components	**$M**	**318**	**351**	**386**	**425**	**463**	**500**	**535**	9

PUERTO RICO – MARKETS (cont)

		1991	1992	1993	1994	1995	1996	1997	
Passive Components									
Capacitors	$M	36	41	43	45	47	48	50	4
Resistors	$M	15	17	18	18	19	19	19	
Connectors	$M	50	43	45	47	49	51	53	4
Relays	$M	27	31	32	32	32	32	33	1
Switches	$M	8	7	7	7	7	7	8	3
Inductors	$M	8	9	9	9	9	10	10	2
Printed Circuit Boards	$M	99	103	110	119	125	129	132	
Total Passive Components	$M	243	251	264	277	288	296	305	4
Other Components									
Microphones, Loudspeakers and Amplifiers	$M	13	15	16	17	17	18	18	4
Aerials	$M	14	13	14	15	16	16	17	6
Unrecorded Media	$M	30	28	29	31	32	33	34	
Accessories & Parts for Consumer Equipment	$M	53	48	50	52	54	56	58	4
Total Other Components	$M	110	104	109	115	119	123	127	4
Total Components Market	$M	671	706	759	817	870	919	967	6

		1991	1992	1993	1994	1995	1996	1997	CAAGR
TOTAL MARKET–PUERTO RICO	$M	1397	1527	1633	1742	1840	1925	2005	6

Notes: 1991 and 1992 are current figures
1993 to 1997 are forecasts at 1992 constant values (ie. inflation is not included)
CAAGR column indicates true annual average growth between 1992 & 1997

4.8 SAUDI ARABIA

BASIC STATISTICS

Land Area:		2 240 000 sq km
Population:		16.9 million (1992)
Civilian Employment:		5.77 million (1990)
of which:	- Agriculture	10%
	- Industry	25% (excluding oil)
	- Other	65%
GDP per Capita:		US$7610 (1992)
TV Sets in Use:		4.5 million
Radio Sets in Use:		5 million
Cars per 1000 Inhabitants:		89 (1992)
Main Telephone Lines per 1000 Inhabitants:		94 (1992)

LEADING ECONOMIC INDICATORS (% increases)

		Actual		Forecast			
	1988	**1989**	**1990**	**1991**	**1992**	**1993**	**1994**
Growth of GDP (Constant market prices)	7.6	0.2	8.4	8.6	5.0	0.8	-3.0
Consumer Price Index	0.9	1.0	2.1	4.7	-0.1	1.8	1.0

Overview

The Kingdom of Saudi Arabia, which occupies over 70% of the Arabian Peninsula, is one of the driest countries in the world. Saudi Arabia, a desert kingdom little more than 25 years ago, is now one of the richest countries in the world and has experienced a phenomenal rate of development. Saudi Arabia has the largest proven oil reserves in the world outside the United States and the Soviet Union. The country incurred considerable costs as a result of the Gulf Conflict and attempted to offset these with increased oil production.

GDP Growth

The production of crude petroleum and petroleum products dominate economic activity and trade activity providing more than 90% of government revenues as well as a substantial proportion of merchandise exports. In consequence, oil price movements and the volume of oil exports have a fundamental impact on economic performance. The increase in oil prices in 1990, as well as higher oil production levels, partly compensating for the loss of Iraqi and Kuwait oil exports, is largely responsible for an acceleration in economic growth of 8.4% and 8.6% in 1990 and 1991 compared with only 0.2% in 1989. Lower levels of oil production in 1992 contained overall GDP growth to 5%. A similar level of growth had been expected in 1993 but lower oil prices and production have resulted in a planned 19% cut in government expenditure. A marginal growth of only 0.8% is now expected with a 3% decline in 1994. Both the public and private sectors are likely to suffer. Within the domestic sector investment spending and government consumption should be reduced given the prospect of delays and cancellations of capital spending programmes and cuts in most government agency budgets. Private capital spending also stands to be hurt by higher real internal rates. Real GDP growth in 1995 is projected as only 1% with the benefits of an expected rise in oil prices partially offset by a decline in production.

Government Policy
Saudi Arabia is an absolute monarchy ruled by King Fahd, who acceded to the thrown in 1982. The king rules in accordance with the Sharia or Islamic law and appoints and heads the Council of Ministers, which serves as an instrument of royal authority in legislative and executive matters. In a move towards evolving a more democratic political structure, King Fahd announced, in February 1992, the formation of a 60 member consultative council, as well as new laws to protect the rights of the individual. The new council will be able to question government decisions and refer any disagreements over policy to the king.

Externally, relations with Yemen and Qatar have been strained recently following separate border disputes with both countries. Talks to try to resolve the disputes can be expected.

Prices & Inflation
Consumer inflation has traditionally been very low and was negative up to 1987. However, the lifting of some subsidies in 1988 and an increase in prices arising from the imposition of more protectionist import tariffs and the worsening of Saudi Arabia's terms of trade, produced minimal inflation in 1988 and 1989. Prices rose more substantially in 1990 to reflect extra costs related to the Gulf crisis. The annual rate of inflation based on Saudi Arabia middle income cost of living de-accelerated to -0.1% in 1992, compared with 4.9% in 1991. The reduction was attributable to the increase in government subsidies on goods & services in March 1992, the rise in local production, and lower import prices as emergency freight insurance premiums were dropped following the Gulf war. Inflation is forecast to remain low in 1993 and 1994.

Industrial Outlook
The rapid development of a Saudi Arabian non-oil industrial sector is a major objective of the whole strategy of successive five year plans. Traditional cottage industries, such as weaving as well as small scale enterprises catering for domestic needs, have always existed in Saudi Arabia, but now all kinds of value-added operations from simple assembly and packaging to metalworking and building materials manufacturing are appearing in the kingdoms newly built industrial estates. Heavy industrial production is almost entirely concentrated in the hands of the state-owned Saudi Arabia Basic Industries Corp. (SABC) but private sector manufacturing has been growing rapidly in scale and range of activities. By 1992, 2316 industrial plants were in operation. These, together with several thousands of unlicensed firms are producing soft drinks, paper products, detergents, furniture, plastics, building products and metal and textile manufacturing. In the next five years most manufacturing projects will be carried out by the private sector and many may be implemented as joint ventures with private or government interests in other Gulf states. The need to control its budget deficit is forcing Saudi Arabia to rationalise its industry strategy. Companies are merging as cost-cutting measures and major construction projects are being delayed.

Labour Market
Total employment in Saudi Arabia increased at an average annual rate of 8% or by 1.42 million between 1980 and 1985, far exceeding the planned annual growth rate of 1.2%. At the end of the period the civilian work force was 4.45 million, the share of Saudis having declined from 49% to 40%. During the fourth plan period (1985-90) the work force grew at an average of 6% p.a. to reach 5.77 million despite projections of a net decline. Under the fifth plan to 1995, annual growth in employment is projected at 0.7%. Employment is targeted to expand at 4.2% p.a.

The 1992 census showed the population to be higher than previous estimates. There were 12.3 million Saudi citizens and 4.6 million non-Saudi residents giving a total population of 16.9 million.

Trade & Balance of Payments
The current account deficit is expected to decline from its very high level of US$19.4 billion in 1992. The trade account is likely to remain in substantial surplus in 1993-94 although below the 1992 level reaching US$ 10.7 billion in 1993 and US$8.7 billion in 1994. In 1993 export revenues are expected to fall to US$39.4 billion because of a 10% decline in oil prices and a reduction of 2% in oil production. In 1994 further reduction in oil prices and non-growth in oil production will restrain revenue to only US$38.3 billion.

ELECTRONICS INDUSTRY

STRUCTURE

Main Production Activities: Computers, Instrumentation, Telecommunications and Communications Equipment.

Major Companies:

- Advanced Electronics Company
- Aldahlawi, M Jamil
- Alireza, Haji Abdullah
- AT&T International
- El Seif, Musaed & Sons
- El-Ajou, Abdul Ghani
- Electric Panel Board Factory Hawa
- Ericsson Saudi Arabia
- Gulf Antenna Factory
- Hawamk Manufacturers
- Hazar Group of Companies
- Metilo Arabia Industrial
- Modern Electronics
- NCR Saudi Arabia
- Raycham Saudi Arabia Ltd
- Raytheon Saudi Arabia
- Remal Electric Factors
- Saudi Arabian-Telecommunications Co (SARTELCO)
- Saudi Electronics & Trading Co (SETCO)
- Saudi Computer Industries (SCI)
- Saudi Amial Est
- The Saudi Lucy Co Ltd
- Wahal Electric Supply Co of Saudi Arabia
- Western Electric International
- Westinghouse Saudi Arabia Ltd

The Saudi Arabian electronics industry is still small and the market is catered for by a large number of trading companies. However, production is increasing in some areas and significant exports are made of computer equipment, instrumentation and measuring equipment and communication equipment.

In July 1993 an exhibition in London included a display by Advanced Electronics Company of Saudi Arabia of aircraft electronics and mobile communications equipment. AEC is the result of one of five projects set up under the Peace Shield offset scheme by the United States.

At the beginning of 1985, a consortium led by Boeing Aerospace was awarded three major contracts worth about US$1.3 billion, and under the terms of the agreement, each contractor had to invest 35% of the value of their contract on high-technology joint ventures in the Kingdom; so that a total of some US$400 million would be invested by the consortium. As part of the qualifying bid, Boeing had to put forward proposals for offset investment projects, and in total they proposed nine, which were all approved by the Offset Committee in the qualifying stage and must be implemented by 1995. Boeing's leading subcontractors are Westinghouse Electric Corporation, ITT Federal Electric, Computer Sciences Corporation (CSC), General Electric Corporation and Frank E Basil. Each member of the consortium had to take responsibility for one or more of the various projects. Included in the list of approved projects were: a US$100 million project to produce electronics equipment for the telecommunications industry for both commercial and military use (Westinghouse); a digital telecommunications factory (ITT); a factory for avionics equipment; a medical products factory (all members); an aerospace maintenance, repair and support centre (Boeing/GEC); and a computer system and services centre (CSC).

The Kingdom's first computer manufacturing plant went into mass production in the middle of 1985. Saved Computer Industries (SCI) was created in response to King Fahd's appeal for the development of national industry and high-technology.

The demand for computer hardware and other electronic products is dependent upon the state of the country's economy, in particular, the level of oil prices. Growth is thus slowing and is forecast to decline in the next year or so as the authorities limit their imports in order to save on hard currency.

The country's Fourth Development Plan (1985-90) allocated SR27.1 million (US$7.2 million) for telecommunications projects, by far the largest share of which is for management, operations and maintenance. The amount did, however, include SR5.1 million for telephone services and SR1.4 million for long-distance communications. The main aims of the fourth plan were to expand the coverage of services and improve their quality, to increase the number of telephone lines and the speed of service, and to increase the use of advanced technology equipment. As a result the Kingdom is on the way to having one of the most modern telephone systems in the world.

The current 5th plan (1990-1995) follows a similar pattern setting a limited number of broad targets for all sectors including the expansion of electricity generators capacity by 25%.

TRADE WITH THE EUROPEAN COMMUNITY

Values in current US$ millions

	Imports to the EC from Saudi Arabia				Exports from the EC to Saudi Arabia			
Product Group	**1989**	**1990**	**1991**	**1992**	**1989**	**1990**	**1991**	**1992**
EDP	1	23	19	30	31	66	81	88
Office Equipment	0	0	0	0	1	3	5	6
Control & Instruments	16	52	48	69	141	181	91	97
Medical	0	0	0	0	12	13	43	38
Industrial	0	0	1	1	9	12	8	9
Communications	21	31	44	56	72	91	127	96
Telecommunications	1	2	1	3	41	30	38	47
Consumer-Video	0	1	1	1	1	9	10	6
Consumer-Audio	0	1	1	2	1	2	2	3
Consumer-Personal	0	0	0	0	5	7	11	12
Active Components	2	2	2	3	1	7	12	9
Passive Components	0	3	4	8	11	31	56	60
Other Components	0	4	39	18	9	15	103	102
TOTAL	41	120	160	191	3354	67	587	573

Quantity in thousands of units

Product Group	Imports to the EC from Saudi Arabia 1990	1991	1992	Exports from the EC to Saudi Arabia 1990	1991	1992
Colour Television	9	0	0	0	1	2
Monochrome Television	0	0	0	0	0	0
Video Cassette Recorders	0	0	0	0	1	2

INDUSTRY DEVELOPMENTS

- AT&T announced in June 1994 that it had won the contract which had been fiercely contested to further expand the telephone network.

- Siemens has built a 10 000 line mobile telephone network and is extending this to 30 000 lines to cover the principal centres in Saudi Arabia at a cost of US$40 million.

- National Telecommunication and telecommunication consultants Salak Ali Abdulrahman have formed a joint venture, TCIL Saudi.

- Ferranti, in May 1994, won a £100 million contract for the Saudi air defence system.

LIST OF SOURCES

Financial Times
Bank Economic Country Reports
Country Report & Country Profile — EIU
International Telecom Statistics — Siemens
World Automotive Statistics — SMMT
Predicasts F&S Index International
Electronic Times
Electronic News
Electronic Business
Electronic Engineering Times
Arab News
Middle East Computing
Merlin
Euromonitor
Central Department of Statistics

SAUDI ARABIA – PRODUCTION

Table No.4.8.1

EDP		1991	1992	1993	1994	CAAGR %
Total EDP Production	SRM	320	330	320	300	-5
	$M	**86**	**88**	**86**	**80**	

OFFICE EQUIPMENT		1991	1992	1993	1994	CAAGR %
Electronic Calculators	SRM	6	6	5	6	0
	$M	**2**	**2**	**1**	**2**	
	No.th	*320*	*320*	*320*	*320*	
Photocopiers	SRM	10	10	10	10	0
	$M	**3**	**3**	**3**	**3**	
Total Office Equipment Production	SRM	16	16	15	16	0
	$M	**4**	**4**	**4**	**4**	

CONTROL & INSTRUMENTATION		1991	1992	1993	1994	CAAGR %
Total Control & Instrumentation	SRM	300	330	340	337	1
	$M	**80**	**88**	**91**	**90**	

MEDICAL & INDUSTRIAL		1991	1992	1993	1994	CAAGR %
Total X-Ray & Medical Equipment	SRM	100	120	132	140	8
	$M	**27**	**32**	**35**	**37**	
Total Industrial Equipment	SRM	90	100	105	109	4
	$M	**24**	**27**	**28**	**29**	
Total Medical & Industrial Production	SRM	190	220	237	249	6
	$M	**51**	**59**	**63**	**67**	

COMMUNICATIONS & RADAR		1991	1992	1993	1994	CAAGR %
Total Communications & Radar	SRM	280	320	340	355	5
	$M	**75**	**86**	**91**	**95**	

TELECOMMUNICATIONS		1991	1992	1993	1994	CAAGR %
Total Telecommunications Production	SRM	160	180	187	190	3
	$M	**43**	**48**	**50**	**51**	

CONSUMER		1991	1992	1993	1994	CAAGR %
Video Equipment						
All Television Sets	SRM	26	44	48	50	7
	$M	**7**	**12**	**13**	**13**	
	No.th	*25*	*40*	*44*	*46*	
Total Video Equipment	SRM	26	44	48	50	7
	$M	**7**	**12**	**13**	**13**	
Audio Equipment						
Radios & Radio Recorders	SRM	13	18	20	21	8
	$M	**3**	**5**	**5**	**6**	
	No.th	*120*	*150*	*170*	*180*	
Total Audio Equipment	SRM	13	18	20	21	8
	$M	**3**	**5**	**5**	**6**	

SAUDI ARABIA – PRODUCTION

Personal Consumer Equipment		**1991**	**1992**	**1993**	**1994**
Total Personal Consumer Equipment	SRM	0	0	0	0
	$M	**0**	**0**	**0**	**0**
Total Consumer Production	SRM	39	62	68	71
	$M	**10**	**17**	**18**	**19**

COMPONENTS		**1991**	**1992**	**1993**	**1994**	CAAGR %
Active Components						
Total Active Components	SRM	0	0	0	0	
	$M	**0**	**0**	**0**	**0**	
Passive Components						
Total Passive Components	SRM	50	60	66	71	9
	$M	**13**	**16**	**18**	**19**	
Other Components						
Unrecorded Media	SRM	30	35	38	40	7
	$M	**8**	**9**	**10**	**11**	
Accessories & Parts for Consumer Equipment	SRM	10	10	10	10	0
	$M	**3**	**3**	**3**	**3**	
Total Other Components	SRM	40	45	48	50	5
	$M	**11**	**12**	**13**	**13**	
Total Components Production	SRM	90	105	114	121	7
	$M	**24**	**28**	**30**	**32**	

		1991	**1992**	**1993**	**1994**	CAAGR %
TOTAL PRODUCTION – SAUDI ARABIA	SRM	1395	1563	1621	1639	2
	$M	**373**	**418**	**433**	**438**	

Notes: 1991 and 1992 are current figures at current exchange rates
1993 & 1994 are forecasts at 1992 constant values & exchange rates (ie inflation is not included)
CAAGR column indicates true annual average growth between 1992 & 1994

Saudi Arabia Trade Balance 1992

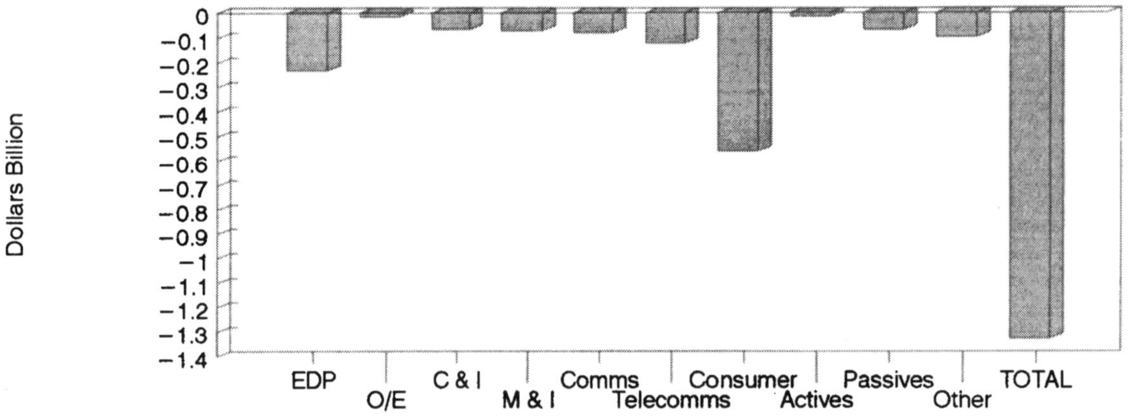

SAUDI ARABIA – MARKETS

Table No.4.8.2

EDP		1991	1992	1993	1994	1995	1996	1997	CAAGR %
Total EDP Market	SRM	1140	1200	1190	1150	1160	1210	1280	1
	$M	**305**	**321**	**318**	**307**	**310**	**324**	**342**	

OFFICE EQUIPMENT		1991	1992	1993	1994	1995	1996	1997	CAAGR %
Electronic Typewriters	SRM	20	17	15	14	13	12	11	−8
	$M	**5**	**5**	**4**	**4**	**3**	**3**	**3**	
	No.th	*20*	*18*	*17*	*17*	*17*	*16*	*16*	
Electronic Calculators	SRM	28	27	26	25	25	26	27	0
	$M	**7**	**7**	**7**	**7**	**7**	**7**	**7**	
	No.th	*470*	*500*	*510*	*520*	*540*	*570*	*600*	
Electronic Cash Registers	SRM	10	11	10	9	9	10	11	0
	$M	**3**	**3**	**3**	**2**	**2**	**3**	**3**	
Photocopiers	SRM	16	18	18	18	20	21	23	5
	$M	**4**	**5**	**5**	**5**	**5**	**6**	**6**	
Total Office Equipment Market	SRM	74	73	69	66	67	69	72	0
	$M	**20**	**20**	**18**	**18**	**18**	**18**	**19**	

CONTROL & INSTRUMENTATION		1991	1992	1993	1994	1995	1996	1997	CAAGR %
Total Control & Instrumentation	SRM	530	580	600	585	590	615	645	2
	$M	**142**	**155**	**160**	**156**	**158**	**164**	**172**	

MEDICAL & INDUSTRIAL		1991	1992	1993	1994	1995	1996	1997	CAAGR %
X–Ray & Medical Equipment									
Total X–Ray & Medical Equipment	SRM	320	320	310	300	305	320	335	1
	$M	**86**	**86**	**83**	**80**	**82**	**86**	**90**	
Total Industrial Equipment	SRM	150	170	175	177	180	186	193	3
	$M	**40**	**45**	**47**	**47**	**48**	**50**	**52**	
Total Medical & Industrial Market	SRM	470	490	485	477	485	506	528	2
	$M	**126**	**131**	**130**	**128**	**130**	**135**	**141**	

COMMUNICATIONS & RADAR		1991	1992	1993	1994	1995	1996	1997	CAAGR %
Total Communications & Radar	SRM	717	623	642	655	668	695	730	3
	$M	**192**	**167**	**172**	**175**	**179**	**186**	**195**	

TELECOMMUNICATIONS		1991	1992	1993	1994	1995	1996	1997	CAAGR %
Total Telecommunications Market	SRM	574	633	646	652	646	633	640	0
	$M	**153**	**169**	**173**	**174**	**173**	**169**	**171**	

SAUDI ARABIA – MARKETS (cont)

CONSUMER		**1991**	**1992**	**1993**	**1994**	**1995**	**1996**	**1997**	CAAGR %
Video Equipment									
All Television Sets	SRM	526	554	529	505	493	507	535	−1
	$M	**141**	**148**	**141**	**135**	**132**	**136**	**143**	
	No.th	*495*	*490*	*470*	*450*	*440*	*450*	*475*	
Video Recorders	SRM	300	270	245	225	220	225	240	−2
	$M	**80**	**72**	**66**	**60**	**59**	**60**	**64**	
	No.th	*320*	*300*	*270*	*250*	*245*	*250*	*265*	
Video Cameras & Camera Combinations	SRM	90	78	74	70	68	72	78	0
	$M	**24**	**21**	**20**	**19**	**18**	**19**	**21**	
	No.th	*38*	*34*	*32*	*30*	*29*	*31*	*34*	
Total Video Equipment	SRM	916	902	848	800	781	804	853	−1
	$M	**245**	**241**	**227**	**214**	**209**	**215**	**228**	
Audio Equipment									
Radios & Radio Combinations	SRM	463	458	430	404	395	403	424	−2
	$M	**124**	**122**	**115**	**108**	**106**	**108**	**113**	
	No.th	*2520*	*2450*	*2300*	*2150*	*2100*	*2150*	*2250*	
Tape Recorders & Players	SRM	110	105	100	95	93	95	100	−1
	$M	**29**	**28**	**27**	**25**	**25**	**25**	**27**	
Record Players	SRM	9	8	7	6	6	6	6	−6
	$M	**2**	**2**	**2**	**2**	**2**	**2**	**2**	
Total Audio Equipment	SRM	582	571	537	505	494	504	530	−1
	$M	**156**	**153**	**144**	**135**	**132**	**135**	**142**	
Personal Consumer Equipment									
Electronic Musical Instruments	SRM	9	9	8	8	8	8	9	0
	$M	**2**	**2**	**2**	**2**	**2**	**2**	**2**	
Electric/Electronic Watches	SRM	650	620	583	548	537	547	575	−1
	$M	**174**	**166**	**156**	**147**	**144**	**146**	**154**	
	No.th	*7700*	*7400*	*7100*	*6800*	*6700*	*6900*	*7400*	
Electric/Electronic Clocks	SRM	53	52	49	46	45	46	48	−2
	$M	**14**	**14**	**13**	**12**	**12**	**12**	**13**	
	No.th	*850*	*800*	*770*	*740*	*730*	*750*	*800*	
Total Personal Consumer Equipment	SRM	712	681	640	602	590	601	632	−1
	$M	**190**	**182**	**171**	**161**	**158**	**161**	**169**	
Total Consumer Market	SRM	2210	2154	2025	1907	1865	1909	2015	−1
	$M	**591**	**576**	**541**	**510**	**499**	**510**	**539**	

COMPONENTS		**1991**	**1992**	**1993**	**1994**	**1995**	**1996**	**1997**	CAAGR %
Active Components									
Total Active Components	SRM	77	55	58	61	67	80	96	12
	$M	**21**	**15**	**16**	**16**	**18**	**21**	**26**	
Passive Components									
Total Passive Components	SRM	280	315	302	278	273	278	286	−2
	$M	**75**	**84**	**81**	**74**	**73**	**74**	**76**	

SAUDI ARABIA – MARKETS (cont)

		1991	1992	1993	1994	1995	1996	1997	
Other Components									
Microphones, Loudspeakers and Amplifiers	SRM	102	109	102	93	90	91	93	–3
	$M	**27**	**29**	**27**	**25**	**24**	**24**	**25**	
Unrecorded Media	SRM	80	87	85	80	80	82	85	0
	$M	**21**	**23**	**23**	**21**	**21**	**22**	**23**	
Accessories & Parts for Consumer Equipment	SRM	177	205	194	177	173	177	183	–2
	$M	**47**	**55**	**52**	**47**	**46**	**47**	**49**	
Total Other Components	SRM	359	401	381	350	343	350	361	–2
	$M	**96**	**107**	**102**	**94**	**92**	**94**	**97**	
Total Components Market	SRM	716	771	741	689	683	708	743	–1
	$M	**191**	**206**	**198**	**184**	**183**	**189**	**199**	

		1991	1992	1993	1994	1995	1996	1997	CAAGR %
TOTAL MARKET – SAUDI ARABIA	SRM	6431	6524	6398	6181	6164	6345	6653	0
	$M	**1720**	**1744**	**1711**	**1653**	**1648**	**1697**	**1779**	

Notes: 1991 and 1992 are current figures at current exchange rates
1993 to 1997 are forecasts at 1992 constant values & exchange rates (ie. inflation is not included)
CAAGR column indicates true annual average growth between 1992 & 1997

4.9 TURKEY

BASIC STATISTICS

Land Area:		777 452 sq km
Population:		58.5 million (1992)
Civilian Employment:		19.2 million (1992)
of which:	- Agriculture	48%
	- Industry	19%
	- Other	33%
GDP per Capita:		US$2108 (1992)
TV Sets in Use:		10.5 million
Radio Sets in Use:		7.1 million
Cars per 1000 Inhabitants:		36 (1992)
Main Telephone Lines per 1000 Inhabitants:		141 (1992)

LEADING ECONOMIC INDICATORS (% increases)

	Actual				Forecast		
	1988	**1989**	**1990**	**1991**	**1992**	**1993**	**1994**
Growth of GDP (constant market prices)	3.8	1.9	9.2	0.3	5.4	7.5	1.5
Consumer Price Index	75.0	65.0	60.0	65.0	70.0	66.0	90.0
Industrial Production	3.1	3.0	8.7	3.1	6.3	9.0	4.0

Overview

Turkey is situated in the extreme south east of Europe, bordering with Iran, Iraq, Syria, Greece, Cyprus and Russia. The country enjoys a relatively stable political situation, with low labour costs, a growing population and high levels of productivity. The monetary unit in Turkey is the Turkish lira.

Turkey belongs to the OECD, IMF, NATO and the Council of Europe. Turkey has had an association agreement with the European Community since 1963, which has full Turkish membership of the European Union as its ultimate aim. However, there have been objections to Turkey joining the EU because of doubts about alleged human rights abuses. In December 1989, the EC announced that it would defer consideration of Turkey's application for full membership until after 1993. However, in April 1992 a trade agreement with EFTA came into operation which guarantees similar concessions to those which exist with the EU.

GNP Growth

Following low growth in 1991 Gross National Product increased by 5.4% in 1992. This growth stemmed mainly from a 9.5% rise in consumption expenditure, while growth in fixed investment expenditure was very mild.

In 1993 the Turkish economy grew at a rate of 6.6% which is above the average long term growth of the economy. Both public sector expenditure and the public sector budget deficit as a share of GNP rose considerably. Also, the real value of credits significantly increased. Private sector savings declined in real terms. The high growth stimulated import demand and exports showed some growth.

Excess liquidity in the market began to rise, reaching very high levels in January 1994 and the Turkish lira depreciated around 50% with very high interest rates and inflation. A sharp decline in the level of economic activity is expected in 1994.

Government Policy

Prime Minister Suleyman Demirel was elected as Head of State by parliament following the death of President Turgut Ozal in April 1993 and was succeeded as premier by Mrs Tansu Ciller, under whose leadership the coalition government of the conservative True Path Party and Social Democrat Populist Party continues.

The government under Mrs Ciller aims to cut the public sector deficit, control inflation and liberalise the economy through privatisation. However, economic reform remains limited. The domestic political situation continues to be overshadowed by increasingly strident disputes with the Kurdish minority in the south east of the country. Turkey, meanwhile, is attempting to build closer economic relations with the Central Asian republics of the former Soviet Union, through improving telecommunications and transport facilities in the region.

Prices & Inflation

Inflation in 1992 increased to an annual average of 70% compared with 65% in 1991. The rate remained high in 1993, the sixth consecutive year of high inflation.

A key factor contributing to the high level of inflation is the persistently large public sector borrowing requirement. In 1992 the PSBR was some US$15.5 billion, equivalent to 12.6% of GNP, compared with a PSBR of 14.4% of GNP in 1991. The central government budget deficit amounted to over US$6 billion in 1992, over two-thirds of which financed subsidies to loss-making state-owned enterprises.

A significant proportion of the PSBR has been financed by advances from the central bank rather than from capital market borrowing. To be successful in curbing inflation, restrictive monetary policy must be supported by an appropriate fiscal policy.

A very significant increase in the inflation rate is expected in 1994.

Industrial Outlook

The last decade has seen Turkey become a predominantly industrialised country. In 1992 manufacturing accounted for nearly 83% of exports compared with an average of around 30% in the years up to 1980. The import substitution policies adopted before 1980 served to provide the country with a well divested economic base but later led to inefficiencies as some industries failed to achieve international competitiveness. Funding the operating losses of the state Economic Enterprises is expected to cost around US$3 billion in 1993. In spite of the general expansion of the private sector, public sector plants still account for about 40% of value added in manufacturing and employ about 20%, or 55 000, of the total industrial workforce.

A legislative framework for privatisation was set up in 1986 but progress has been slow. The PTT is planned for privatisation but in 1993 preliminary moves were blocked by the constitutional court.

Labour Market

Despite persistently high inflation, real wages increased in 1992 and this was partly responsible for lack of progress in reducing unemployment, which in 1993 was estimated at around 12.5% of the workforce. High population growth has also contributed to the high rate of unemployment.

The problem is compounded by the population shift from rural to urban areas and further complicated by the large number of informal participants in the agricultural sector. In the past the migration of Turkish workers to Western Europe, especially West Germany and the Middle East, alleviated the problem. However, changes in both regions have decreased their demand for migrant workers.

Trade & Balance of Payments
A sharp rise in imports and a decline in official transfers to Turkey was largely responsible for a balance of payments current account deficit of almost US$1 billion in 1992, compared with a surplus of US$272 million in 1991. Merchandise exports increased by almost 9% to around US$14.9 billion in 1992, while imports rose by 10% to US$23.1 billion.

The structure of the balance of payments has changed little in recent years. The trade balance was in deficit by some US$8.2 billion in 1992, while there was a significant surplus on both the invisibles balance (excluding interest payments) and on transfer payments.

High imports in 1993 are expected to result in a sustained increase in the current account deficit to at least US$5 billion, falling again in 1994 to around US$1 billion.

ELECTRONICS INDUSTRY

STRUCTURE

Number of Major Electronics Companies (1990):

Components	23
Consumer	17
Telecommunications	21
Defence	2
Medical, Industrial, Instruments, Communications	27
TOTAL (excl EDP)	90

Main Production Activities: Consumer Equipment (mainly colour televisions), Telecommunications & Communications Equipment

Major Companies

	Component	Consumer	Telecommunications	Defence	EDP
AFA	x				
Aselsan	x		x	x	
Bekoteknik		x			
Cihan		x			
Denge	x				
Disan	x				
Ede	x				
Elak	x				

	Component	Consumer	Telecommunications	Defence	EDP
Elsa	x	x	x		
Ensa	x				
Erdem	x	x			
Ericsson		x			
Fuze		x			
Ies		x			
Imper		x			
Geoteknik			x		
Hes Kablo			x		
Hopsan	x				
IBM					x
Iska-Rabak			x		
Izoper	x				
Karel			x		
Kondas	x				
Karel (Vestel)		x			
Kavel Kablo			x		
Koc-Unisys					x
Kumtel			x		
Meta		x			
Milteknik		x			
Neta		x	x		
Nevtron		x			
Netas	x				
NCR					x
Ortas			x		
Panc	x				
Ptt	x		x		
Raks		x			
Savronik				x	
Simko			x		
Simtel		x			
Sun	x				
Teleks			x		
Telesas	x				
Teletas	x		x		
Telkom	x				
Telra	x	x			
Termal	x				
Test				x	
Testas	x		x		
Tunik	x				
Tupko			x		

	Component	Consumer	Telecommunications	Defence	EDP
Turk Kablo			x		
Turk Philips	x	x			
Turk Siemens			x		
Turk Telkas			x		
Unal	x				
Vestel	x	x			
Yenicag		x			
Zer	x				

Electronics Industry Employment (1990):

Components	2230
Consumer	12915
Telecommunications	9074
Defence	1239
Medical, industrial, instruments, communications	3885
TOTAL (excl EDP)	29343

The Turkish electronics industry exhibited fast growth during the 1980s but in 1992 overall growth was minimal, due mainly to a decline of 10% in colour television production. EC imports from Turkey increased by 160% in 1990 but by only 21% in 1991 and fell by 2% in 1992. The majority of these imports were of colour televisions with consumer video products accounting for 73% in 1992.

High demand from the former Comecom bloc, which was satisfied by Western European countries partly procured from Turkey, aided export growth but this demand has now subsided. The domestic market also slowed in 1992. A more buoyant domestic market occurred in 1993 but is forecast to decline in 1994 although improved conditions in the export market should enable production to show some growth.

The start of the electronics industry in Turkey was around 1958 when the Quota Regime banned the import of electronic consumer goods. In 1964 a decree for the Assembly Industry served as a regulating factor and the National Security Council initiated studies for the establishment of the electronics industry. The local manufacture of black and white TV receivers commenced in 1968. TRT, the Turkish Radio and Television Broadcasting Organisation, started regular TV broadcasting in 1972 and passed to colour in 1983.

Production of black and white TV sets peaked in the mid-1970s but had dropped considerably when colour production started in 1982. Increasing competition from imported products resulted in improvements in production techniques and large-scale exports started in 1987, especially to EC countries.

VCR production commenced in 1984 and increased rapidly with 230 000 units being produced in 1986. Domestic demand reduced with the advent of more TV channels and producers failed to keep up with technical developments. A similar situation existed with audio equipment, with production reaching a peak in 1986 and then reducing rapidly because of imports and economic conditions, then increasing slightly in 1989 as exports increased.

In Turkey, the use of cash registers was enforced in 1985 resulting in a buoyant market. At present six companies are manufacturing.

The production of microwave ovens started in 1988 in a joint venture with Goldstar. More than 90% of production is exported.

The Turkish telecommunications industry commenced in 1965 with the establishment of a research laboratory by the PTT as PPT ARLA, later to become Teletas. In 1967 Netas was set up jointly by the PTT and Northern Telecom of Canada to manufacture exchanges and telephone sets. Other companies established in the 1960s included TTE, Simko, Teleks, Kumtel and Telesis. In 1985 Teletas signed a business agreement with Bell Telephone Manufacturing Co. to manufacture system 12 digital exchanges. Teletas also manufactured radio links, multiplex systems and optical fibre transmission systems, partly under licence and partly by its own developments. This has resulted in 50% digitalization by 1990. The first vehicle telephone system came into use in 1986 and it is planned to convert to GSM system in 1992. In 1988 the PTT slowed its investment and went into international bidding. This resulted in a drop in demand for the Turkish telecommunications industry and a decrease of about 3000 in the number of employees. PTT orders increased in 1990 and with higher exports, particularly to East Europe countries, production expanded in 1992.

Black and white television tubes are now only produced by Ensa, and although Vestel and Videocolour of the French Thompson Group obtained permission to build a colour TV tube factory in 1990, the project was stopped. Carbon film and metal oxide resistors and ceramic capacitors are being manufactured by Testas and film capacitors by Testap, Tupko, Ede and Kondas, meeting part of the domestic needs. Printed circuit boards are produced by a number of small workshops and by the large equipment manufacturers for their own use, but most of the laminate is imported.

Turkey has various ASIC design centres in operation and the necessary infrastructure and technology for the production of prototypes and pilot projects provided by the Tubitak Marmara Research Center Yital, however there is no large-scale semiconductor industry.

Almost all coils and transformers and some connectors are produced locally but the Turkish components industry remains very small.

Other electronic products produced included broadcast and studio equipment by Bes, Era, Zer and Enesis, factory automation and test and measuring equipment and medical electronics equipment. Karel started manufacturing personal computers in 1990 and a personal computer developed by Tubitak Industries is being bought to a commercial state. In 1990 Vestel began making monitors in large quantities for export. There are around 1880 establishments selling computers, producing software, designing systems and producing computers from imported piece parts. The top company is IBM followed by KOC Unisys and NCR.

TRADE WITH THE EUROPEAN COMMUNITY

Values in current US$ millions

	Imports to the EC from Turkey				Exports from the EC to Turkey			
Product Group	**1989**	**1990**	**1991**	**1992**	**1989**	**1990**	**1991**	**1992**
EDP	1	27	18	16	90	129	141	139
Office Equipment	0	0	0	0	0	6	4	6
Control & Instruments	0	4	5	7	55	97	105	115
Medical	0	0	1	1	13	30	27	25
Industrial	0	0	0	0	6	5	9	16
Communications	9	9	4	5	30	73	35	42
Telecommunications	2	1	2	2	68	69	93	114
Consumer-Video	60	173	213	208	18	26	26	30
Consumer-Audio	20	21	12	5	7	20	14	10
Consumer-Personal	0	0	1	8	9	4	6	
Active Components	0	1	2	2	26	44	50	61
Passive Components	1	5	6	10	11	36	82	83
Other Components	0	0	28	29	4	16	62	16
TOTAL	93	241	291	286	336	560	652	663

Quantity in thousands of units

Product Group	**Imports to the EC from m**			**Exports from the Turkey EC to Turkey**		
	1990	**1991**	**1992**	**1990**	**1991**	**1992**
Colour Television	39	35	37	872	954	959
Monochrome Television	1	1	2	2	0	0
Video Cassette Recorders	3	23	8	0	6	5

LIST OF SOURCES

Financial Times
Bank Economic Country Reports
Country Report & Country Profile — EIU
OECD Economic Survey
OECD Economic Outlook
Garanti Bank
Euromonitor
International Telecom Statistics — Siemens
World Automotive Statistics — SMMT
Predicasts F&S Index International
Electronic Times
Electronic News
Electronic Business
Electronic Engineering Times
State Institute of Statistics
Almanac Turkish Electronics Industry — Electronic Industralists Association
Economic Forecasts - Elsevier

TURKEY – PRODUCTION

Table No.4.9.1

EDP		1991	1992	1993	1994	CAAGR %
Total EDP Production	TLBn	584	996	1100	1127	
	$M	**140**	**145**	**160**	**164**	6
OFFICE EQUIPMENT		**1991**	**1992**	**1993**	**1994**	CAAGR %
Electronic Calculators	TLBn	13	27	27	27	
	$M	**3**	**4**	**4**	**4**	0
	No.th	*900*	1200	1300	1300	
Electronic Cash Registers	TLBn	92	131	124	124	
	$M	**22**	**19**	**18**	**18**	-3
	No.th	*63*	59	57	58	
Photocopiers	TLBn	29	48	48	48	
	$M	**7**	**7**	**7**	**7**	0
	No.th	*13*	*13*	*13*	*13*	
Total Office Equipment Production	TLBn	134	206	199	199	
	$M	**32**	**30**	**29**	**29**	-2
CONTROL & INSTRUMENTATION		**1991**	**1992**	**1993**	**1994**	CAAGR %
Total Control & Instrumentation	TLBn	17	69	76	76	
	$M	**4**	**10**	**11**	**11**	5
MEDICAL & INDUSTRIAL		**1991**	**1992**	**1993**	**1994**	CAAGR %
Total X-Ray & Medical Equipment	TLBn	209	412	474	495	
	$M	**50**	**60**	**69**	**72**	10
Total Industrial Equipment	TLBn	33	55	62	62	
	$M	**8**	**8**	**9**	**9**	6
Total Medical & Industrial Production	TLBn	242	467	536	557	
	$M	**58**	**68**	**78**	**81**	9
COMMUNICATIONS & RADAR		**1991**	**1992**	**1993**	**1994**	CAAGR %
Total Communications & Radar	TLBn	334	687	825	859	
	$M	**80**	**100**	**120**	**125**	12
TELECOMMUNICATIONS		**1991**	**1992**	**1993**	**1994**	CAAGR %
Exchange Equipment	TLBn	709	1491	1526	1560	
	$M	**170**	**217**	**222**	**227**	2
Other Equipment & Parts	TLBn	501	825	838	852	
	$M	**120**	**120**	**122**	**124**	2
Total Telecommunications Production	TLBn	1210	2316	2364	2412	
	$M	**290**	**337**	**344**	**351**	2

TURKEY – PRODUCTION (cont)

CONSUMER		1991	1992	1993	1994	
Video Equipment						
Colour Television	TLBn	4522	6783	7161	7415	
	$M	**1084**	**987**	**1042**	**1079**	5
	No.th	*2300*	*2185*	*2350*	*2450*	
Video Recorders	TLBn	129	137	103	69	
	$M	**31**	**20**	**15**	**10**	-29
	No.th	*71*	*47*	*37*	*25*	
Total Video Equipment	TLBn	4652	6920	7264	7484	
	$M	**1115**	**1007**	**1057**	**1089**	4
Audio Equipment						
Total Audio Equipment	TLBn	375	653	687	715	
	$M	**90**	**95**	**100**	**104**	5
Personal Consumer Equipment						
Electronic Musical Instruments	TLBn	4	7	7	7	
	$M	**1**	**1**	**1**	**1**	0
	No.th	*0*	*0*	*0*	*0*	
Electric/Electronic Watches	TLBn	38	55	62	62	
	$M	**9**	**8**	**9**	**9**	6
	No.th	*2300*	*2000*	*2200*	*2200*	
Electric/Electronic Clocks	TLBn	13	14	14	14	
	$M	**3**	**2**	**2**	**2**	0
	No.th	*900*	*600*	*620*	*610*	
Total Personal Consumer Equipment	TLBn	54	76	82	82	
	$M	**13**	**11**	**12**	**12**	4
Total Consumer Production	TLBn	5081	7649	8033	8281	
	$M	**1218**	**1113**	**1169**	**1205**	4

COMPONENTS		1991	1992	1993	1994	CAAGR %
Active Components						
Television Tubes	TLBn	17	27	34	34	
	$M	**4**	**4**	**5**	**5**	12
Other Valves & Tubes	TLBn	13	27	27	27	
	$M	**3**	**4**	**4**	**4**	0
Diodes	TLBn	8	21	21	21	
	$M	**2**	**3**	**3**	**3**	0
Other Discrete Semiconductors	TLBn	4	21	27	27	
	$M	**1**	**3**	**4**	**4**	15
ICs & Other Microcircuits	TLBn	17	41	48	55	
	$M	**4**	**6**	**7**	**8**	15
Total Active Components	TLBn	58	137	158	165	
	$M	**14**	**20**	**23**	**24**	10
Passive Components						
Capacitors	TLBn	13	69	76	82	
	$M	**3**	**10**	**11**	**12**	10
Resistors	TLBn	4	7	7	7	
	$M	**1**	**1**	**1**	**1**	0
Connectors & Switches	TLBn	33	69	82	82	
	$M	**8**	**10**	**12**	**12**	10

TURKEY – PRODUCTION (cont)

		1991	1992	1993	1994	
Transformers & Reactors	TLBn	125	241	254	254	
	$M	**30**	**35**	**37**	**37**	3
Relays	TLBn	17	34	41	34	
	$M	**4**	**5**	**6**	**5**	0
Printed Circuit Boards	TLBn	50	103	117	131	
	$M	**12**	**15**	**17**	**19**	13
Total Passive Components	TLBn	242	522	577	591	
	$M	**58**	**76**	**84**	**86**	6
Other Components						
Loudspeakers	TLBn	13	21	21	21	
	$M	**3**	**3**	**3**	**3**	0
Amplifiers	TLBn	4	7	7	7	
	$M	**1**	**1**	**1**	**1**	0
Aerials	TLBn	33	62	69	76	
	$M	**8**	**9**	**10**	**11**	11
Unrecorded Media	TLBn	242	454	474	488	
	$M	**58**	**66**	**69**	**71**	4
Accessories & Parts for Consumer Equipment	TLBn	63	103	110	110	
	$M	**15**	**15**	**16**	**16**	3
Total Other Components	TLBn	355	646	680	701	
	$M	**85**	**94**	**99**	**102**	4
Total Components Production	TLBn	655	1306	1416	1457	
	$M	**157**	**190**	**206**	**212**	6

		1991	1992	1993	1994	CAAGR %
TOTAL PRODUCTION–TURKEY	TLBn	8256	13696	14548	14967	
	$M	**1979**	**1993**	**2117**	**2178**	5

Notes: 1991 and 1992 are current figures at current exchange rates
1993 & 1994 are forecasts at 1992 constant values & exchange rates (ie inflation is not included)
Current estimates & forecasts are made in US$ with local values shown for guidance only
CAAGR column indicates true annual average growth between 1992 & 1994

Turkey Trade Balance 1992

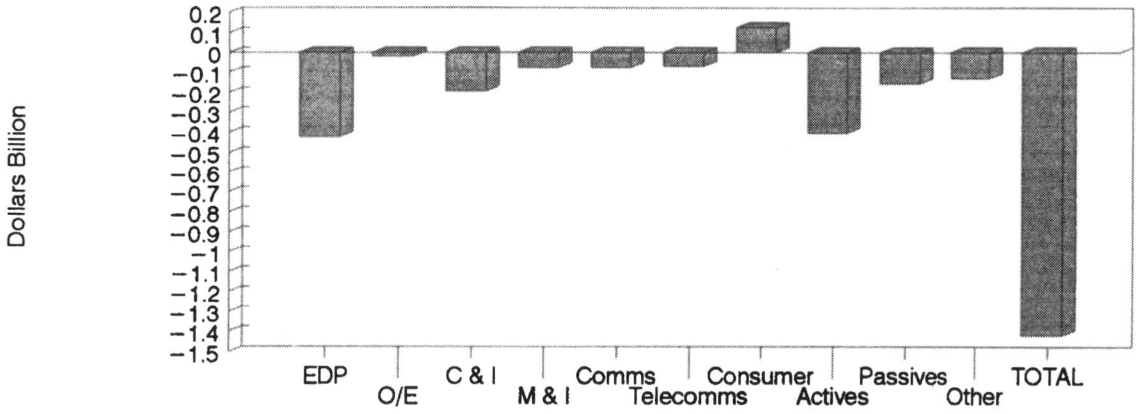

TURKEY – MARKETS

Table No.4.9.2

EDP		**1991**	**1992**	**1993**	**1994**	**1995**	**1996**	**1997**	CAAGR %
Total EDP Market	TLBn	2220	3883	4274	4185	4350	4659	4982	
	$M	**532**	**565**	**622**	**609**	**633**	**678**	**725**	5
OFFICE EQUIPMENT		**1991**	**1992**	**1993**	**1994**	**1995**	**1996**	**1997**	CAAGR %
Electronic Typewriters	TLBn	17	14	14	14	14	14	14	
	$M	**4**	**2**	**2**	**2**	**2**	**2**	**2**	0
	No.th	*60*	*40*	*41*	*41*	*42*	*43*	*44*	
Electronic Calculators	TLBn	25	41	41	41	41	41	41	
	$M	**6**	**6**	**6**	**6**	**6**	**6**	**6**	0
	No.th	*1500*	*1600*	*1650*	*1620*	*1650*	*1700*	*1750*	
Cash Registers	TLBn	88	124	124	124	124	124	131	
	$M	**21**	**18**	**18**	**18**	**18**	**18**	**19**	1
	No.th	*60*	*56*	*58*	*57*	*58*	*59*	*60*	
Photocopiers	TLBn	88	158	165	158	165	172	172	
	$M	**21**	**23**	**24**	**23**	**24**	**25**	**25**	2
	No.th	*31*	*33*	*35*	*34*	*36*	*38*	*39*	
Total Office Equipment Market	TLBn	217	337	344	337	344	350	357	
	$M	**52**	**49**	**50**	**49**	**50**	**51**	**52**	1
CONTROL & INSTRUMENTATION		**1991**	**1992**	**1993**	**1994**	**1995**	**1996**	**1997**	CAAGR %
Total Control & Instrumentation	TLBn	918	1402	1471	1443	1484	1574	1670	
	$M	**220**	**204**	**214**	**210**	**216**	**229**	**243**	4
MEDICAL & INDUSTRIAL		**1991**	**1992**	**1993**	**1994**	**1995**	**1996**	**1997**	CAAGR %
Total X–Ray & Medical Equipment	TLBn	401	735	811	838	886	948	1010	
	$M	**96**	**107**	**118**	**122**	**129**	**138**	**147**	7
Total Industrial Equipment	TLBn	138	247	268	282	295	316	337	
	$M	**33**	**36**	**39**	**41**	**43**	**46**	**49**	6
Total Medical & Industrial Market	TLBn	538	983	1079	1120	1182	1264	1347	
	$M	**129**	**143**	**157**	**163**	**172**	**184**	**196**	7
COMMUNICATIONS & RADAR		**1991**	**1992**	**1993**	**1994**	**1995**	**1996**	**1997**	CAAGR %
Total Communications & Radar	TLBn	784	1209	1306	1333	1416	1498	1601	
	$M	**188**	**176**	**190**	**194**	**206**	**218**	**233**	6
TELECOMMUNICATIONS		**1991**	**1992**	**1993**	**1994**	**1995**	**1996**	**1997**	CAAGR %
Exchange Equipment	TLBn	943	1691	1725	1642	1663	1697	1732	
	$M	**226**	**246**	**251**	**239**	**242**	**247**	**252**	0
Other Equipment & Parts	TLBn	634	1106	1127	1100	1106	1127	1148	
	$M	**152**	**161**	**164**	**160**	**161**	**164**	**167**	1
Total Telecommunications Market	TLBn	1577	2797	2852	2742	2769	2824	2879	
	$M	**378**	**407**	**415**	**399**	**403**	**411**	**419**	1

TURKEY – MARKETS (cont)

CONSUMER		1991	1992	1993	1994	1995	1996	1997	CAAGR %
Video Equipment									
Colour Television	TLBn	3617	5566	5834	5539	5642	5862	6095	
	$M	**867**	**810**	**849**	**806**	**821**	**853**	**887**	2
	No.th	*1350*	*1410*	*1490*	*1460*	*1500*	*1580*	*1650*	
Monochrome Television	TLBn	4	7	7	7	7	7	7	
	$M	**1**	**1**	**1**	**1**	**1**	**1**	**1**	0
	No.th	*25*	*20*	*20*	*19*	*19*	*19*	*20*	
Video Recorders	TLBn	167	186	199	192	199	206	213	
	$M	**40**	**27**	**29**	**28**	**29**	**30**	**31**	3
	No.th	*90*	*62*	*67*	*65*	*68*	*71*	*74*	
Video Cameras & Camera Combinations	TLBn	25	34	41	41	48	55	62	
	$M	**6**	**5**	**6**	**6**	**7**	**8**	**9**	12
	No.th	*10*	*9*	*10*	*11*	*13*	*15*	*17*	
Total Video Equipment	TLBn	3813	5793	6082	5779	5896	6130	6377	
	$M	**914**	**843**	**885**	**841**	**858**	**892**	**928**	2
Audio Equipment									
Total Audio Equipment	TLBn	409	722	756	722	735	763	797	
	$M	**98**	**105**	**110**	**105**	**107**	**111**	**116**	2
Personal Consumer Equipment									
Electronic Musical Instruments	TLBn	25	34	34	34	34	34	34	
	$M	**6**	**5**	**5**	**5**	**5**	**5**	**5**	0
	No.th	*0*	*0*	*0*	*0*	*0*	*0*	*0*	
Electric/Electronic Watches	TLBn	121	213	220	213	220	227	234	
	$M	**29**	**31**	**32**	**31**	**32**	**33**	**34**	2
	No.th	*8300*	*8000*	*8300*	*8100*	*8400*	*8600*	*8800*	
Electric/Electronic Clocks	TLBn	17	34	34	34	34	34	41	
	$M	**4**	**5**	**5**	**5**	**5**	**5**	**6**	4
	No.th	*1400*	*1800*	*1830*	*1810*	*1850*	*1870*	*2000*	
Total Personal Equipment	TLBn	163	282	289	282	289	295	309	
	$M	**39**	**41**	**42**	**41**	**42**	**43**	**45**	2
Total Consumer Market	TLBn	4385	6796	7126	6783	6920	7188	7484	
	$M	**1051**	**989**	**1037**	**987**	**1007**	**1046**	**1089**	2

COMPONENTS		1991	1992	1993	1994	1995	1996	1997	CAAGR %
Active Components									
Television Tubes	TLBn	893	1319	1409	1477	1546	1649	1752	
	$M	**214**	**192**	**205**	**215**	**225**	**240**	**255**	6
Other Valves & Tubes	TLBn	50	82	82	82	82	89	89	
	$M	**12**	**12**	**12**	**12**	**12**	**13**	**13**	2
Diodes	TLBn	79	137	144	144	151	158	165	
	$M	**19**	**20**	**21**	**21**	**22**	**23**	**24**	4
Transistors	TLBn	96	144	144	144	144	151	158	
	$M	**23**	**21**	**21**	**21**	**21**	**22**	**23**	2
Other Discrete Semiconductors	TLBn	96	144	151	151	158	165	172	
	$M	**23**	**21**	**22**	**22**	**23**	**24**	**25**	4
ICs & Other Microcircuits	TLBn	601	1072	1113	1134	1148	1209	1326	
	$M	**144**	**156**	**162**	**165**	**167**	**176**	**193**	4
Total Active Components	TLBn	1815	2900	3044	3134	3230	3422	3663	
	$M	**435**	**422**	**443**	**456**	**470**	**498**	**533**	5

TURKEY – MARKETS (cont)

		1991	1992	1993	1994	1995	1996	1997	CAAGR
Passive Components									
Capacitors	TLBn	217	316	330	337	344	357	371	
	$M	**52**	**46**	**48**	**49**	**50**	**52**	**54**	3
Resistors	TLBn	92	144	151	151	151	151	158	
	$M	**22**	**21**	**22**	**22**	**22**	**22**	**23**	2
Connectors & Switches	TLBn	321	550	570	577	591	605	625	
	$M	**77**	**80**	**83**	**84**	**86**	**88**	**91**	3
Transformers & Inductors	TLBn	196	261	268	268	268	275	282	
	$M	**47**	**38**	**39**	**39**	**39**	**40**	**41**	2
Relays	TLBn	63	124	124	124	124	131	137	
	$M	**15**	**18**	**18**	**18**	**18**	**19**	**20**	2
Printed Circuit Boards	TLBn	108	192	206	213	220	234	247	
	$M	**26**	**28**	**30**	**31**	**32**	**34**	**36**	5
Total Passive Components	TLBn	997	1587	1649	1670	1697	1752	1821	
	$M	**239**	**231**	**240**	**243**	**247**	**255**	**265**	3
Other Components									
Microphones	TLBn	8	21	21	21	21	21	21	
	$M	**2**	**3**	**3**	**3**	**3**	**3**	**3**	0
Loudspeakers	TLBn	63	137	144	144	151	158	165	
	$M	**15**	**20**	**21**	**21**	**22**	**23**	**24**	4
Amplifiers	TLBn	21	27	27	27	27	34	34	
	$M	**5**	**4**	**4**	**4**	**4**	**5**	**5**	5
Aerials	TLBn	58	96	103	110	117	124	131	
	$M	**14**	**14**	**15**	**16**	**17**	**18**	**19**	6
Unrecorded Media	TLBn	217	371	385	392	399	412	433	
	$M	**52**	**54**	**56**	**57**	**58**	**60**	**63**	3
Accessories & Parts for Consumer	TLBn	772	886	921	921	935	948	983	
Equipment	**$M**	**185**	**129**	**134**	**134**	**136**	**138**	**143**	2
Total Other Components	TLBn	1139	1539	1601	1615	1649	1697	1766	
	$M	**273**	**224**	**233**	**235**	**240**	**247**	**257**	3
Total Components Market	TLBn	3951	6027	6295	6418	6577	6872	7250	
	$M	**947**	**877**	**916**	**934**	**957**	**1000**	**1055**	4

		1991	1992	1993	1994	1995	1996	1997	CAAGR
TOTAL MARKET – TURKEY	TLBn	14589	23434	24746	24361	25042	26230	27570	
	$M	**3497**	**3410**	**3601**	**3545**	**3644**	**3817**	**4012**	3

Notes: 1991 and 1992 are current figures at current exchange rates
1993 to 1997 are forecasts at 1992 constant values & exchange rates (ie. inflation is not included)
Current estimates & forecasts are made in US$ with local values shown for guidance only
CAAGR column indicates true annual average growth between 1992 & 1997

4.10 VENEZUELA

BASIC STATISTICS

Land Area:		912 000 sq km
Population:		20.3 (1992)
Civilian Employment:		7.4 million (1991)
of which:	- Agriculture	12%
	- Manufacturing	27%
	- Other	61%
GDP per Capita:		US$3017 (1992)
TV Sets in Use:		3.7 million
Radio Sets in Use:		8.1 million
Cars per 1000 Inhabitants:		74 (1992)
Main Telephone Lines per 1000 Inhabitants:		86 (1992)

LEADING ECONOMIC INDICATORS (% increases)

	Actual				Forecast		
	1988	**1989**	**1990**	**1991**	**1992**	**1993**	**1994**
Growth of GDP (Constant Market Price)	6.2	-7.8	6.9	10.4	7.3	-1.0	-1.0
Consumer Price Index	36.6	81.0	36.5	31.0	31.9	40.0	35.0
Industrial Production (non oil)	3.4	-14.5	5.0	10.4	9.9	na	na

Overview
Venezuela is located in the north of South America bordering on the west with Colombia, on the east with Guyana and on the south with Brazil. The country has one of the most stable political systems in South America, since 1958 all governments have been civilians, democratically elected. Venezuela has vast petroleum and mineral reserves but since the early 1980s the economy has been badly hit by depressed oil prices. However, petroleum still accounted for 79% of exports in 1992. The monetary unit in Venezuela is the bolivar.

GDP Growth
Between 1990 and 1992, Venezuela enjoyed the highest rate of economic growth in Latin America, with real Gross Domestic Product growth averaging 8.2% per annum. This impressive performance largely reflected the success of the government's economic policies, as well as an extensive public investment programme in the petroleum sector and higher oil output during the Gulf crisis. However, in the first six months of 1993, real GDP growth fell to just 2.8%, compared with the same period a year earlier. This was partly because of political insecurity as well as the introduction of more restrictive monetary policies by the central bank, which mainly took the form of sharp increases in interest rates. With the political situation worsening the economy further slowed with negative growth of around 1% forecast for 1993. The first months of 1994 saw a fall in production and sales across a range of sectors. This is likely to continue throughout 1994 as domestic demand remains subdued.

Government Policy
The Social Democratic Acción Democrátical (AD) government took office for a five year

term in February 19890 and introduced an economic adjustment programme. However, subsequent austerity measures implemented by the government proved unpopular and there followed serious social unrest.

On 21 May 1993 President Carlos Andrés Pérez was impeached by the senate in the wake of the Supreme Court charging him with the misappropriation and embezzlement of state funds. Against this background, the Congress elected Ramón J Velásquez as interim president.

The "Ley Habiltante" or Enabling Act, passed in August 1993 granted the interim government special decree powers. These have given the government greater scope to continue with its economic reform programme. The decrees passed so far have mainly comprised urgent measures aimed at reducing the public sector deficit and regenerating the privatisation programme.

In February 1994 Rafael Caldera was inaugerated as president and in his first speech spoke of his desire to head a government that was both popular and neo-liberal. He also issued a decree suspending the so called "economic guarentee".

Inflation & Prices
Inflation increased to 81% in 1989, following the easing of price controls. Following the correction in relative prices, inflation checked significantly in 1990. However, with demand pressures building up there was little further progress in lowering inflation in 1991 with the increase in consumer prices remaining well above 30%. A continued widening of the public sector deficit has again increased inflation with consumer prices rising by 37.3% in the twelve months to August 1993, compared with 31.9% in the twelve months to December 1992.

Industrial Outlook
The oil industry is the mainstay of the economy, having generated 20.4% of GDP and almost four-fifths of export earnings in 1992. However, value-added in the oil sector declined by 1.9% in 1992 which contrasted sharply with the 10.3% increase recorded in 1991. Meanwhile, the non-oil sector grew by 9.9% in 1992, compared with 10.4%in the previous year. Manufacturing, the second most important sector to oil, with a contribution of 17.5% to GDP, expanded by 12.1% compared with an increase of 10.8% in 1991. Construction activity, which generates 6.4% of GDP also grew by 16.8% in 1992, compared with 30.8% in 1991. Finance and property, which combined account for 13.3% of GDP, expanded by 8.6% last year, compared with 4.9% in 1991.Strongest growth however, was registered by trade and tourism, which accounted for 14.6% of GDP, expanded by 17%, compared with 8.7% in 1991.

Trade & Balance of Payments
Oil revenues fell by 13% in 1991 to US$12.1 billion, as average earnings per barrel fell by 23% to US$15.70, primarily because of weaker international demand and lower world oil prices. Non-oil exports also declined, by 20% to US$2.8 billion in 1991, while imports surged by 49% to US$11billion. Non-oil exports rose only slightly to just under US$3billion, while imports increased to US$12.3 billion. As a result, the trade surplus narrowed for a second successive year to US$1.7 billion, while imports increased to US$ 12.3 billion. As a result, the trade surplus narrowed for a second successive year to US$1.7 billion. Reflecting

trends in merchandise trade, the current account moved into deficit by US$3.4 billion, or the equivalent of 5.6% of GDP.

An improvement is forecast for 1993 and 1994 as oil reserves increase and imports fall in line with the depressed economy. The trade balance is forecast as US$3.2 billion in 1993 and US$3.8 billion in 1994 with current account deficits of US$1.8 billion and US$0.9 billion respectively.

ELECTRONIC INDUSTRY

STRUCTURE

Main Production Activities: Computer Peripherals, Instrumentation, Telecommunications Equipment, Communication Equipment, Tape Recorders, Printed Circuit Boards.

Major Companies:

AETI
Auto Radio
C A Electrolux
CANTV
Dispositivos de Telecomunicaciones Venezolanas (DISTELCA)
Distribuidoras Unidas
Electrologica del Centro (ELECTROCEN)
Electronica y Telecommunicaciones (EYT)
Electronica del Centro (ECCA)
Fonolab
Industrias Electronicas (INDELEC)
Industrias Regina
Industrias
General Electric de Venezuela
Hewlett Packard de Venezuela
IBM Co de Venezuela SA
Microtel Electronica
National de Venezuela
Philips
Tele-Norma
Televisores Venezolanos SA (TELEVENSA)
Voltex

Due to of a lack of its own state-of-the-art technology, scarcity of skilled cheap labour, lack of basic research in electronics, international competition, the initial set-up costs and too small a market for local products to allow for economies of scale, as is the case for chips, Venezuela has been highly dependent on imports for almost all of its electronic goods. However, the Venezuelan government is encouraging a local electronics industry, to save some of the US$983 million which was spent in 1992 on imports of various types of electronic equipment and components. The priorities, according to the Director General (Technology) in the Ministry of Development, are computer software, telecommunications equipment, computer hardware, electronic components and electronic consumer goods.

In 1990, electronics production expanded by a real 24% with the emphasis on computer equipment, instrumentation, communications equipment and passive components, mainly printed circuit boards. In 1991 real production growth was 14.6% followed by further growth in 1992 of 7.3%. Growth is forecast to continue at over 5% per annum in spite of a reduction in the domestic market. However, total output in 1994 will only be US$574 million with most of the estimated market of US$1450 being satisfied by imports.

The Venezuelan government were proposing to introduce a highly protectionist regime of import barriers and tariff protection, which would have virtually excluded all foreign investment and technology. However, the government decided to relax the highly restrictive interpretation of Andean Pact regulations on foreign investment and exempt electronics, information technology and biotechnology. Whilst reserving the field of telecommunications to the national government, therefore, the way was open for interested foreign investors to become involved in the electronics and computers industries in Venezuela.

With the encouragement of the Ministry of Development, there are already a number of firms in existence which import components from a number of different sources and assemble them into systems which they market under their own name. This trend is likely to continue, although it is improbable that Venezuela will ever develop a sophisticated industry of its own.

CANTV is the state-owned telephone company and the dominant company in the telecommunications field.

The multinational presence in Venezuelan electronics includes GTE, General Electric, Siemens, Ericsson, Thomson-CSF (mainly in the area of telecommunications); Philips, Sony, National, and Pioneer (mostly consumer products).

TRADE WITH THE EUROPEAN COMMUNITY

Values in current US$ millions

	Imports to the EC from Venezuela				Exports from the EC to Venezuela			
Product Group	**1989**	**1990**	**1991**	**1992**	**1989**	**1990**	**1991**	**1992**
EDP	0	0	0	0	17	15	12	23
Office Equipment	0	0	0	0	0	1	2	2
Control & Instruments	4	1	0	1	19	47	23	21
Medical	0	0	0	0	2	2	4	5
Industrial	0	0	0	0	2	1	3	3
Communications	1	1	1	2	5	15	20	17
Telecommunications	0	0	0	0	21	10	17	75
Consumer-Video	0	0	0	0	0	0	0	1
Consumer-Audio	0	0	0	0	0	0	1	0
Consumer-Personal	0	0	0	0	0	0	1	1
Active Components	0	0	0	0	0	2	2	6
Passive Components	0	0	1	0	7	9	21	18
Other Components	0	0	1	0	0	2	7	4
TOTAL	5	2	3	3	73	104	113	176

LIST OF SOURCES

Financial Times
Bank Economic Country Reports
Country Report & Country Profile — EIU
International Telecom Statistics — Siemens
World Automotive Statistics — SMMT
Predicasts F&S Index International
Electronic Times
Electronic News
Electronic Business
Electronic Engineering Times
Euromonitor
Veneconomy Monthly
Business Latin America
Banco Central of Venezuela
Annuario del Comercio Exterior

VENEZUELA – PRODUCTION

Table No.4.10.1

EDP		1991	1992	1993	1994	CAAGR %
Total EDP Production	BrM	7384	10944	12038	13270	
	$M	**130**	**160**	**176**	**194**	10

OFFICE EQUIPMENT		1991	1992	1993	1994	CAAGR %
Electronic Typewriters	BrM	568	684	684	684	
	$M	**10**	**10**	**10**	**10**	0
Electronic Calculators	BrM	284	342	342	342	
	$M	**5**	**5**	**5**	**5**	0
Electronic Cash Registers	BrM	114	68	68	137	
	$M	**2**	**1**	**1**	**2**	41
Total Office Equipment Production	BrM	966	1094	1094	1163	3
	$M	**17**	**16**	**16**	**17**	

CONTROL & INSTRUMENTATION		1991	1992	1993	1994	CAAGR %
Total Control & Instrumentation	BrM	1988	2736	2873	3010	
	$M	**35**	**40**	**42**	**44**	5

MEDICAL & INDUSTRIAL		1991	1992	1993	1994	CAAGR %
Medical Equipment						
Total X–Ray & Medical Equipment	BrM	1420	1915	2052	2189	
	$M	**25**	**28**	**30**	**32**	7
Industrial Equipment						
Total Industrial Equipment	BrM	1250	1642	1710	1778	
	$M	**22**	**24**	**25**	**26**	4
Total Medical & Industrial Production	BrM	2670	3557	3762	3967	
	$M	**47**	**52**	**55**	**58**	6

COMMUNICATIONS & RADAR		1991	1992	1993	1994	CAAGR %
Total Communications & Radar	BrM	2670	3557	3762	3830	
	$M	**47**	**52**	**55**	**56**	4

TELECOMMUNICATIONS		1991	1992	1993	1994	CAAGR %
Total Telecommunications Production	BrM	3408	3420	3625	3694	
	$M	**60**	**50**	**53**	**54**	4

VENEZUELA – PRODUCTION (cont)

CONSUMER		1991	1992	1993	1994	
Video Equipment						
Colour Television	BrM	1477	2052	2326	2462	
	$M	**26**	**30**	**34**	**36**	10
	No.th	*110*	*125*	*140*	*150*	
Monochrome Television	BrM	682	1094	1163	1163	
	$M	**12**	**16**	**17**	**17**	3
	No.th	*190*	*250*	*270*	*280*	
Total Video Equipment	BrM	2158	3146	3488	3625	
	$M	**38**	**46**	**51**	**53**	7
Audio Equipment						
Radios & Radio Combinations	BrM	1022	1368	1436	1436	
	$M	**18**	**20**	**21**	**21**	2
	No.th	*500*	*550*	*580*	*590*	
Tape Recorders & Decks	BrM	852	1026	1026	1026	
	$M	**15**	**15**	**15**	**15**	0
	No.th	*400*	*400*	*400*	*400*	
Record Players	BrM	57	68	68	68	
	$M	**1**	**1**	**1**	**1**	0
	No.th	*30*	*30*	*29*	*28*	
Total Audio Equipment	BrM	1931	2462	2531	2531	1
	$M	**34**	**36**	**37**	**37**	
Personal Consumer Equipment						
Electric/Electronic Watches	BrM	568	616	616	684	
	$M	**10**	**9**	**9**	**10**	11
	No.th	*2200*	*2000*	*2100*	*2200*	
Electric/Electronic Clocks	BrM	511	616	616	616	
	$M	**9**	**9**	**9**	**9**	0
	No.th	*2000*	*1900*	*1950*	*2000*	
Total Personal Consumer Equipment		1079	1231	1231	1300	
	$M	**19**	**18**	**18**	**19**	3
Total Consumer Production	BrM	5169	6840	7250	7456	
	$M	**91**	**100**	**106**	**109**	4

COMPONENTS		1991	1992	1993	1994	CAAGR %
Active Components						
Discretes, Diodes & Transistors	BrM	114	274	342	410	
	$M	**2**	**4**	**5**	**6**	22
Total Active Components	BrM	114	274	342	410	
	$M	**2**	**4**	**5**	**6**	

VENEZUELA – PRODUCTION (cont)

		1991	1992	1993	1994	
Passive Components						
Capacitors	BrM	227	274	274	274	
	$M	**4**	**4**	**4**	**4**	0
Connectors	BrM	227	342	410	410	
	$M	**4**	**5**	**6**	**6**	10
Relays	BrM	170	205	205	205	
	$M	**3**	**3**	**3**	**3**	0
Inductors	BrM	284	410	410	410	
	$M	**5**	**6**	**6**	**6**	0
Printed Circuit Boards	BrM	682	889	889	958	
	$M	**12**	**13**	**13**	**14**	4
Total Passive Components	BrM	1590	2120	2189	2257	
	$M	**28**	**31**	**32**	**33**	3
Other Components						
Amplifiers	BrM	114	68	68	68	
	$M	**2**	**1**	**1**	**1**	0
Accessories & Parts for Consumer Equipment	BrM	341	342	342	342	
	$M	**6**	**5**	**5**	**5**	0
Total Other Components	BrM	454	410	410	410	
	$M	**8**	**6**	**6**	**6**	0
Total Components Production	BrM	2158	2804	2941	3078	
	$M	**38**	**41**	**43**	**45**	5

		1991	1992	1993	1994	CAAGR %
TOTAL PRODUCTION–VENEZUELA	BrM	26412	34952	37346	39467	
	$M	**465**	**511**	**546**	**577**	6

Notes: 1991 and 1992 are current figures at current exchange rates
1993 & 1994 are forecasts at 1992 constant values & exchange rates (ie inflation is not included)
Current estimates & forecasts are made in US$ with local values shown for guidance only
CAAGR column indicates true annual average growth between 1992 & 1994

Venezuela Trade Balance 1992

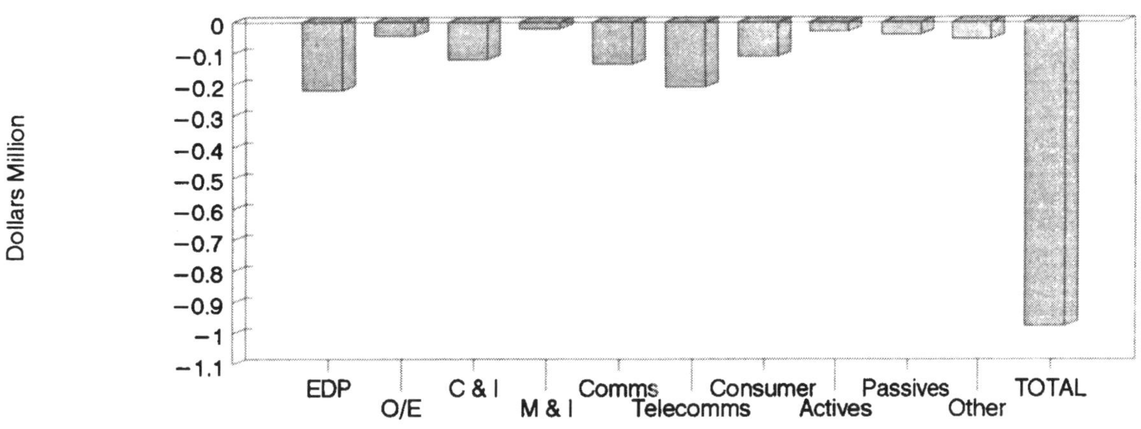

VENEZUELA – MARKETS

Table No.4.10.2

EDP		1991	1992	1993	1994	1995	1996	1997	CAAGR %
Total EDP Market	BrM	20164	25924	25308	25650	27702	30438	33516	
	$M	**355**	**379**	**370**	**375**	**405**	**445**	**490**	5

OFFICE EQUIPMENT		1991	1992	1993	1994	1995	1996	1997	CAAGR %
Electronic Typewriters	BrM	1022	1300	1231	1163	1094	1026	1026	
	$M	**18**	**19**	**18**	**17**	**16**	**15**	**15**	−5
Electronic Calculators	BrM	568	752	752	752	821	821	821	
	$M	**10**	**11**	**11**	**11**	**12**	**12**	**12**	2
Electronic Cash Registers	BrM	227	342	342	410	410	479	479	
	$M	**4**	**5**	**5**	**6**	**6**	**7**	**7**	7
Photocopiers	BrM	1250	1642	1778	1847	1984	2120	2257	
	$M	**22**	**24**	**26**	**27**	**29**	**31**	**33**	7
Total Office Equipment Market	BrM	3067	4036	4104	4172	4309	4446	4583	
	$M	**54**	**59**	**60**	**61**	**63**	**65**	**67**	3

CONTROL & INSTRUMENTATION		1991	1992	1993	1994	1995	1996	1997	CAAGR %
Total Control & Instrumentation	BrM	9542	10876	10328	10123	10534	11149	11833	
	$M	**168**	**159**	**151**	**148**	**154**	**163**	**173**	2

MEDICAL & INDUSTRIAL		1991	1992	1993	1994	1995	1996	1997	CAAGR %
Medical Equipment									
Total X–Ray & Medical Equipment	BrM	2158	2873	3010	3146	3420	3762	4172	
	$M	**38**	**42**	**44**	**46**	**50**	**55**	**61**	8
Industrial Equipment									
Total Industrial Equipment	BrM	1931	2120	2120	2120	2120	2326	2394	
	$M	**34**	**31**	**31**	**31**	**31**	**34**	**35**	2
Total Medical & Industrial Market	BrM	4090	4993	5130	5267	5540	6088	6566	
	$M	**72**	**73**	**75**	**77**	**81**	**89**	**96**	6

COMMUNICATIONS & RADAR		1991	1992	1993	1994	1995	1996	1997	CAAGR %
Total Communications & Radar	BrM	6873	12722	12449	12244	12586	13201	13885	
	$M	**121**	**186**	**182**	**179**	**184**	**193**	**203**	2

TELECOMMUNICATIONS		1991	1992	1993	1994	1995	1996	1997	CAAGR %
Total Telecommunications Market	BrM	6646	17716	15937	14364	13612	13885	14159	
	$M	**117**	**259**	**233**	**210**	**199**	**203**	**207**	−4

VENEZUELA – MARKETS (cont)

CONSUMER		1991	1992	1993	1994	1995	1996	1997	CAAGR %
Video Equipment									
Colour Television	BrM	2783	4446	4720	5062	5404	5882	6361	
	$M	**49**	**65**	**69**	**74**	**79**	**86**	**93**	7
	No.th	*200*	*255*	*270*	*290*	*310*	*330*	*350*	
Monochrome Television	BrM	852	1368	1300	1231	1231	1163	1231	
	$M	**15**	**20**	**19**	**18**	**18**	**17**	**18**	–2
	No.th	*250*	*330*	*320*	*310*	*300*	*290*	*280*	
Video Recorders	BrM	341	958	1026	1094	1163	1300	1436	
	$M	**6**	**14**	**15**	**16**	**17**	**19**	**21**	8
	No.th	*24*	*40*	*44*	*48*	*52*	*58*	*64*	
Video Cameras	BrM	341	616	684	752	821	958	1026	
	$M	**6**	**9**	**10**	**11**	**12**	**14**	**15**	11
	No.th	*10*	*15*	*17*	*19*	*21*	*23*	*25*	
Total Video Equipment	BrM	4317	7387	7729	8140	8618	9302	10055	
	$M	**76**	**108**	**113**	**119**	**126**	**136**	**147**	6
Audio Equipment									
Radios & Radio Combinations	BrM	1704	2462	2668	2804	2941	3146	3352	
	$M	**30**	**36**	**39**	**41**	**43**	**46**	**49**	6
	No.th	*1000*	*1150*	*1250*	*1320*	*1400*	*1500*	*1600*	
Car Radios	BrM	227	342	342	410	479	547	616	
	$M	**4**	**5**	**5**	**6**	**7**	**8**	**9**	12
	No.th	*60*	*70*	*75*	*85*	*95*	*105*	*115*	
Tape Recorders & Decks	BrM	1193	1710	1778	1847	1915	1984	2052	
	$M	**21**	**25**	**26**	**27**	**28**	**29**	**30**	4
	No.th	*530*	*590*	*610*	*630*	*650*	*670*	*700*	
Record Players & Decks	BrM	284	342	342	274	274	274	274	
	$M	**5**	**5**	**5**	**4**	**4**	**4**	**4**	–4
	No.th	*130*	*130*	*125*	*120*	*115*	*110*	*100*	
Total Audio Equipment	BrM	3408	4856	5130	5335	5609	5951	6293	
	$M	**60**	**71**	**75**	**78**	**82**	**87**	**92**	5
Personal Consumer Equipment									
Electric/Electronic Watches	BrM	909	1300	1368	1436	1505	1573	1642	
	$M	**16**	**19**	**20**	**21**	**22**	**23**	**24**	5
	No.th	*5200*	*6000*	*6300*	*6500*	*6700*	*7000*	*7200*	
Electric/Electronic Clocks	BrM	568	752	821	821	889	889	958	
	$M	**10**	**11**	**12**	**12**	**13**	**13**	**14**	5
	No.th	*2240*	*2200*	*2300*	*2400*	*2500*	*2600*	*2700*	
Total Personal Equipment	BrM	1477	2052	2189	2257	2394	2462	2599	
	$M	**26**	**30**	**32**	**33**	**35**	**36**	**38**	5
Total Consumer Market	BrM	9202	14296	15048	15732	16621	17716	18947	
	$M	**162**	**209**	**220**	**230**	**243**	**259**	**277**	6

COMPONENTS		1991	1992	1993	1994	1995	1996	1997	CAAGR %
Active Components									
Television Tubes	BrM	454	684	752	821	889	889	958	
	$M	**8**	**10**	**11**	**12**	**13**	**13**	**14**	7
Other Valves & Tubes	BrM	114	68	68	68	68	68	68	
	$M	**2**	**1**	**1**	**1**	**1**	**1**	**1**	0

VENEZUELA – MARKETS (cont)

		1991	1992	1993	1994	1995	1996	1997	
Diodes & Transistors	BrM	398	479	479	479	547	616	684	
	$M	**7**	**7**	**7**	**7**	**8**	**9**	**10**	7
Other Discrete Semiconductors	BrM	114	205	205	205	205	274	274	
	$M	**2**	**3**	**3**	**3**	**3**	**4**	**4**	6
ICs & Other Microcircuits	BrM	682	889	958	1026	1163	1300	1505	
	$M	**12**	**13**	**14**	**15**	**17**	**19**	**22**	11
Total Active Components	BrM	1761	2326	2462	2599	2873	3146	3488	
	$M	**31**	**34**	**36**	**38**	**42**	**46**	**51**	8
Passive Components									
Capacitors	BrM	454	684	752	752	821	821	889	
	$M	**8**	**10**	**11**	**11**	**12**	**12**	**13**	5
Resistors	BrM	284	274	274	274	274	342	342	
	$M	**5**	**4**	**4**	**4**	**4**	**5**	**5**	5
Connectors	BrM	1079	1436	1505	1573	1642	1710	1847	
	$M	**19**	**21**	**22**	**23**	**24**	**25**	**27**	5
Inductors	BrM	454	547	547	616	616	616	684	
	$M	**8**	**8**	**8**	**9**	**9**	**9**	**10**	5
Relays	BrM	170	205	205	205	205	205	205	
	$M	**3**	**3**	**3**	**3**	**3**	**3**	**3**	0
Switches	BrM	511	547	547	547	547	616	616	
	$M	**9**	**8**	**8**	**8**	**8**	**9**	**9**	2
Printed Circuit Boards	BrM	795	1094	1163	1231	1300	1436	1505	
	$M	**14**	**16**	**17**	**18**	**19**	**21**	**22**	7
Total Passive Components	BrM	3749	4788	4993	5198	5404	5746	6088	
	$M	**66**	**70**	**73**	**76**	**79**	**84**	**89**	5
Other Components									
Microphones and Loudspeakers	BrM	227	410	410	410	410	410	410	
	$M	**4**	**6**	**6**	**6**	**6**	**6**	**6**	0
Amplifiers	BrM	227	342	342	342	342	342	342	
	$M	**4**	**5**	**5**	**5**	**5**	**5**	**5**	0
Aerials	BrM	852	1368	1368	1300	1368	1436	1436	
	$M	**15**	**20**	**20**	**19**	**20**	**21**	**21**	1
Unrecorded Media	BrM	568	1026	958	958	958	1026	1094	
	$M	**10**	**15**	**14**	**14**	**14**	**15**	**16**	1
Accessories & Parts for Consumer Equipment	BrM	738	958	958	958	958	1026	1094	
	$M	**13**	**14**	**14**	**14**	**14**	**15**	**16**	3
Total Other Components	BrM	**2613**	**4104**	**4036**	**3967**	**4036**	**4241**	**4378**	
	$M	**46**	**60**	**59**	**58**	**59**	**62**	**64**	1
Total Components Market	BrM	8122	11218	11491	11765	12312	13133	13954	
	$M	**143**	**164**	**168**	**172**	**180**	**192**	**204**	4

		1991	1992	1993	1994	1995	1996	1997	CAAGR
TOTAL MARKET – VENEZUELA	BrM	67706	101779	99796	99317	103216	110056	117443	
	$M	**1192**	**1488**	**1459**	**1452**	**1509**	**1609**	**1717**	3

Notes: 1991 and 1992 are current figures at current exchange rates
1993 to 1997 are forecasts at 1992 constant values & exchange rates (ie. inflation is not included)
Current estimates & forecasts are made in US$ with local values shown for guidance only
CAAGR column indicates true annual average growth between 1992 & 1997

APPENDIX 5.1

EXCHANGE RATES

Exchange rates against the US dollar (Annual Average)

Country	Currency		1987	1988	1989	1990	1991	1992
AUSTRALIA	Australian Dollar	A$	1.43	1.28	1.26	1.28	1.28	1.36
AUSTRIA	Schilling	Sch	12.60	12.30	13.20	11.37	11.68	11.0
BELGIUM	Franc	BF	37.3	36.8	39.4	33.4	34.1	32.1
BRAZIL	Cruzerio	Crz	39.2	262	2863	68.8	406.6	4502
CANADA	Canadian Dollar	C$	1.33	1.23	1.18	1.17	1.15	1.21
DENMARK	Krone	DKr	6.84	6.73	7.31	6.19	6.40	6.05
FINLAND	Markka	Fmk	4.40	4.18	4.29	3.82	4.04	4.48
FRANCE	Franc	FFr	6.01	5.96	6.38	5.45	5.64	5.29
GERMANY	Deutsche Mark	DM	1.80	1.76	1.88	1.62	1.66	1.56
HONG KONG	Hong Kong Dollar	HK$	7.8	7.8	7.8	7.8	7.8	7.74
INDIA	Rupee	Rup	13.0	13.9	16.2	17.5	22.7	25.9
INDONESIA	Rupiah	Rp	1644	1686	1770	1843	1950	2030
IRELAND	Irish Pound	I£	0.67	0.66	0.70	0.60	0.62	0.59
ISRAEL	Shekel	Shk	1.60	1.60	1.92	2.02	2.28	2.46
ITALY	Lira	Lira	1296	1302	1372	1198	1241	1232
JAPAN	Yen	¥	145	128	138	145	135	127
MALAYSIA	Ringgit	Rt	2.52	2.62	2.71	2.70	2.75	2.57
NETHERLANDS	Guilder	Gd	2.03	1.98	2.12	1.82	1.87	1.76
NORWAY	Krone	NKr	6.74	6.52	6.90	6.26	6.48	6.21
PHILIPPINES	Philippine Peso	Peso	20.60	21.10	21.74	24.31	27.48	25.5
SINGAPORE	Singapore Dollar	S$	2.11	2.01	1.95	1.81	1.73	1.63
SOUTH AFRICA	Rand	Rd	2.04	2.26	2.62	2.58	2.76	2.80
SOUTH KOREA	Won	Wn	823	731	671	708	733	781
SPAIN	Peseta	Pts	123.0	116.0	118.4	101.9	103.9	102.4
SWEDEN	Krona	Swkr	6.34	6.13	6.45	5.92	6.05	5.82
SWITZERLAND	Franc	SFr	1.49	1.46	1.64	1.39	1.43	1.41
TAIWAN	Taiwan Dollar	NT$	31.40	28.59	26.41	26.67	26.83	25.0
THAILAND	Baht	Bt	25.70	25.30	25.70	25.59	25.52	25.4
UNITED KINGDOM	Pound Sterling	£	0.61	0.56	0.61	0.56	0.57	0.57
EUROPEAN COMMUNITY	ECU	0.87	0.85	0.91	0.79	0.81	0.77	
CHINA	Yuan	Yn	3.72	3.72	3.76	4.78	5.32	5.52
EGYPT	Egyptian Pound	E£	1.52	0.7	1.1	2.0	3.33	3.33
GREECE	Drachma	Dr	135.4	141.9	162.4	158.5	182.3	190.6
MEXICO	Peso	Ps	1.38	2.3	2.5	2.8	3.0	3.1
NEW ZEALAND	N Z Dollar	NZ$	1.69	1.52	1.67	1.68	1.73	1.86
PORTUGAL	Escudo	Esc	141.0	144.0	157.5	142.6	144.5	135.0
SAUDI ARABIA	Riyal	SR	3.74	3.74	3.74	3.74	3.74	3.74
TURKEY	Turkish Lira	TL	857	1422	2122	2609	4172	6872
VENEZUELA	Boliva	Br	14.5	14.5	34.7	46.9	56.8	68.4

Source: IMF

APPENDIX 5.2

GUIDE TO THE INTERPRETATION OF THE STATISTICS

This section assesses the main technical factors which can distort comparison of Yearbook estimates of electronics activity with other sources and defines the key terms used in the Yearbook to assist interpretation.

Production
The total output of products fabricated by indigenous manufacturers in a country for sale both for home consumption or for export, but excluding imports and re-exports of finished goods. Some manufacture may amount to little more than assembly of imported piece-parts and sub-assemblies; however, the most widely accepted definition allows such products to be included as production, provided value is added in the assembly process and the finished article can be classified under a different SIC (Standard Industrial Classification) from those of its components.

Where imports of such parts are significant, adjustment is made in calculating the market by the number of items (units) ignoring the formula "Production + Imports - Exports = Market" to avoid double counting.

Most production statistics record sales of products and thus exclude goods produced in-house for use by the company itself in its original establishment. However, products shipped overseas are included in import and export statistics with a deduced value being applied, if necessary, and would thus normally be included in production figures. US production statistics record shipments of all products sold or transferred to other establishments of the same company with an appropriate allocation of company overhead and profit.

If no production is shown for a product in a particular country, it may be assumed that the market for the product is serviced by importers only.

Market
The total volume of product consumed by end-users in the country during a particular year, whether shipped by indigenous manufacturers or imported, but excluding exports or re-exports. Yearbook markets are largely estimates of Apparent Consumption based on the calculation:

Production + Imports - Exports = Apparent Consumption

unless lack of data, or unreliable source material, precludes the application of this formula, in which case the estimate is based on appropriate alternative sources (such as desk research, industry estimates, company annual reports, consultancy visit reports, input/output analysis, etc.)

Consumption of items produced in-house and not shipped outside the establishment or group of establishments are regarded as closed markets and are not included.

Imports & Exports
Most countries now use the Harmonized System for trade statistics which simplifies international comparison. Appendix 5.3 gives the codes used in the compilation of the

Yearbook, but for simplicity, these are only shown to the 4 digit level. Trade is actually analyzed to the 8th, 9th or even 10th digit level with over 350 categories being employed.

It should be noted that exports shown on the summary tables include re-exports. Thus, countries with significant entrepot activities, notably Hong Kong and Singapore, but also some European countries such as Belgium and the Netherlands, have higher than expected imports, part of which are then re-exported.

Products

The products and groups of products included in the Yearbook are chosen, as far as possible, to reflect the usage of electronics. Electronic equipment is fairly easily defined, but some components, particularly passive components, are more difficult to separate from electrical and even mechanical types. To enable international comparison between countries, which is one of the main aims of the Yearbook, broad product headings are standardised. However, the exact coverage of some sectors may vary due to the non-availability of source data. This is a particular problem with relays, switches, inductors and connectors which may include some electrical products in addition to the electronic types.

Military Equipment

In the past the bulk of military equipment was composed of radar and navigational aids and included in the product group **'Communications and Military'**. However, increasingly, most types of electronic equipment, except consumer, contain some which is used for military purposes. To more accurately describe the Yearbook coverage the product group has been renamed **'Communications & Radar'** from the 1994 edition of the Yearbook. Whilst this will continue to include radar and navigational aids used for military purposes, it must be appreciated that other types of military equipment and components are included in other product headings. Where it is possible to separately identify military equipment this will be named as such and listed under the 'Communications and Radar' heading.

Forecasts

Each edition of the Yearbook gives estimates of markets and production from the base data 2 years before the date of the edition. Later figures are forecasts made at constant exchange rates using estimated real growth, that is inflation and exchange rate fluctuations are not included.

These forecasts relate to future growth rates for existing types of products or new products which will fall into the same statistical classification as existing products. Wholly innovative products not available at the base year and outside current statistical classifications, but which may be introduced later and obtain a significant sales volume within the forecast period, are beyond the scope of the Yearbook and are not included.

The main factors taken into account when forecasting are the historical performance of the series in question, the general economic outlook and the major growth influencing factors for each sector, which we have considered as follows:

Electronic Capital Equipment:	industrial investment outlook public sector purchasing policy technical change product pricing trends

	government controls (e.g. import restrictions, taxation) maturity of the industry
Electronic Consumer Equipment:	consumer expenditure outlook technical change product pricing trends government controls (e.g. credit, taxation, trade restrictions) maturity of the industry (e.g. ownership saturation levels)
Electronic Components:	demand in all other end-equipment sectors technical change product pricing trends captive/merchant market situation government controls (e.g. import restrictions) maturity of the industry

In addition, knowledge of published company activities (such as major contracts, plant expansion or shut-down programmes, military and industrial development projects, technical and price trends), much of which information is now summarized in our annual publication, the World Electronics Companies File, is also given due weight as applicable.

Growth Rates
For both Production and Market data, Compound Annual Average Growth Rates (CAAGR) are given for each product based on the local currency data (see note on exchange rates below).

The CAAGR reflects the annual average growth rate for each individual product between the base year and the last year given, that is two years for production and 5 years for markets.

This will enable readers to identify, with relative ease, the important areas of high or low growth within the electronics industry, although this may vary appreciably over the period.

Exchange Rates
For the majority of the countries covered in the Yearbook, values are given in local currency with the value in US dollars shown to facilitate international comparison. Constant exchange rates are used for the forecast years but the current rate is used for the historical data. Year-on-year fluctuations in exchange rates can be significant, with the result that a negative trend revealed in local currency values can turn into an upswing when expressed in US dollars, or vice versa, and for this reason it is inappropriate to calculate growth rates, including comparisons of the volume of electronics' business in various countries, from the US dollar series (see note on Growth Rates above).

For some of the Newly Industrialized Countries (NICs) and Emerging Nations and all of the East European countries where exchange rates fluctuate dramatically and most trade is conducted in dollars, values have been expressed in US dollars and forecasts made in this currency. In these cases, local currency values have been calculated from the US dollar values but these serve as a guide only and consideration of the volume of the electronics business and of growth should be taken from the US dollar values for these countries.

Inflation

Forecast figures indicate real growth, that is, excluding inflation. However, historical data show current figures and therefore include inflation. This can result in an apparently higher growth rate for earlier years and must be noted when considering long term trends. An indication of real growth is usually given in the overview in the summary but for more detailed examination of real growths we recommend the Yearbook data on diskette (see Subscriber Enquiry Service at the end of this section).

Prices

Production values are at ex-factory prices net of discounts, allowances, freight charges and returns. Import values are CIF, (including carriage, insurance and freight costs to the port of final destination) and export values are FOB (free-on board). Import duty, VAT and other local taxes, and importer and distributor margins are not included.

Market values therefore represent the above and do not include distributors or retailers margins. In the case of consumer equipment these are very significant and can be the reason for apparent inconsistencies between Yearbook estimates and other sources of data.

Stocks Levels

Changes in stock levels may affect calculations, particularly in the consumer equipment sector. Consumer equipment markets are usually estimated from manufacturers' deliveries to the home market plus imports and are not, therefore, measurements of retail sales (the consumer off-take), which may vary from estimates shown according to wholesale and retail stock levels.

Production and market statistics also include manufacturers' inventories, not usually a significant proportion except occasionally in component statistics. Wherever possible, manufacturers' shipments (i.e. production excluding inventories) have been used to calculate apparent consumption data. In certain other instances, production is recorded as invoiced sales of home produced goods and should not, therefore, include manufacturers' stocks.

Where shipment data are not available, the Production + Imports - Exports = Apparent Consumption formula can become particularly distorted in the highly volatile trading conditions often found, for example, in SE Asia, where manufacturers producing in large quantities for overseas markets may suddenly find themselves left with large stocks as a result of rapid changes in demand. These may be stored until better times and exported in a different year to that in which they were manufactured. As far as possible, a correction is made for this type of effect in our market data when thought to be significant, and can result in an estimate being made for production or trade which differs from an 'official' figure.

Accessories, Parts and Spare Parts

Accessories, parts and spare parts are included as a separate item wherever possible, mainly for comparative purposes, since in a number of countries a separate estimate of this heading is not available although its value is included in total shipments of equipment.

Subscriber Enquiry Service

As part of the Yearbook subscription, subscribers are entitled to use the Subscriber Enquiry Service with regard to any queries relating to the present and historical statistical series or any other aspect of this Yearbook. In particular, arrangements can be made to supply the forecast date at an alternative exchange rate. This would be taken from the 'Yearbook on Diskette' series which has been revised to a similar format to the book.

The research team also welcomes suggestions for improving the coverage of the Yearbook or supplying other information needs of subscribers on a periodical basis.

All enquiries should be addressed directly to Elsevier at the address below:

Elsevier Advanced Technology
The Boulevard
Langford Lane
Kidlington
Oxford OX5 1GB, United Kingdom
Tel: (+44)(0)865 843848/843000
Fax: (+44)(0)865 843971

APPENDIX 5.3
A GUIDE TO STATISTICAL TRADE CLASSIFICATIONS

In 1988 the Harmonized System was introduced for import and export statistics which is now used by the majority of industrialised countries. The following table gives the codes (to the 4 digit level) together with the related Standard Indusrtial Trade Classification (SITC) used in the compilation of the Yearbook.

	HS	SITC(R3)
EDP	8471	752.1
		752.2
		752.3
		752.6
		752.7
		752.9
	8473	759.9
Office Equipment	8469	751.1
	8470	751.2
	8519	763.3
		763.8
	8520	763.8
	9009	751.3
		759.1
Control &	8543	778.7
Instrumentation	9014	874.1
	9015	874.1
	9023	874.5
	9024	874.5
	9025	874.5
	9026	874.3
	9027	874.4
	9030	874.7
	9031	874.2
	9032	874.6
X-Ray & Medical	9018	774.1
Equipment	9021	899.6
	9022	774.2
Industrial Equipment	8514	741.3
	8530	778.8
	8531	778.8
	9013	871.9
Communications &	8525	764.3
Military	8526	764.8

	8527	764.8
	8529	764.9
	9014	874.1
	9015	874.1
Telecommunication	8517	764.1
	8517	764.9
	8520	763.8
Consumer Video	8521	763.8
	8525	764.8
	8528	761.1
		761.2
Consumer Audio	8519	763.3
	8519	763.8
	8520	763.8
	8527	762.1
	8527	762.2
	8527	762.8
Consumer Personal	9006	881.1
	9101	885.3
	9102	885.4
	9103	885.7
	9105	885.7
	9207	898.2
Active Components	8540	776.2
	8541	776.3
		776.8
	8542	776.4
Passive Components	8504	771.1
		771.2
	8532	778.6
	8533	772.3
	8534	772.2
	8535	772.4
	8536	772.5
Other Components	8518	764.2
		764.9
	8522	764.9
	8523	898.4
	8528	761.1
	8529	764.9

APPENDIX 5.4 GUIDE TO THE DEFINITION OF THE ELECTRONICS PRODUCT HEADINGS

ELECTRONIC DATA PROCESSING

Computers and Complete Systems	Analogue and hybrid/digital data processing machines. Digital data processing machines with any capacity of random access memory, combined or not combined with an input and output unit. Covers main frame, mini, micro and home computers including those computers used as word processors.
Peripherals Storage	Storage units including central storage units, disc storage units, magnetic tape storage units.
Input/Output	Input or Output units including printers, preparation equipment, readers, punches, visual display units, multiplexers and modems used for computer links, other terminals and consoles.
Accessories and Parts	Parts, accessories and sub-assemblies for use solely or principally with machines as above. This includes all remaining EDP equipment not fully covered by the above definitions. Complete computers assembled by retailers are included.
Software	This is not included except where it constitutes a part of a complete system.

OFFICE EQUIPMENT

Electronic Typewriters	Electric and electronic typewriters, including portables. Including automatic typewriters, typewriters with a built-in memory and dedicated word processors except those which can also be used as a personal computer.
Electronic Calculators	Electronic calculating machines including desk and hand-held types, with or without a printing device.
Electronic Cash Registers	Electronic cash registers incorporating a calculating device including POS terminals that are not connected to a computer.

Electronic Accounting Machines	Electronic accounting machines all types including book-keeping and invoicing machines.
Dictation Equipment	Desk and hand-held dictating machines of audio office type including transcription machines.
Photocopiers	Thermo-copying apparatus, contact-type photocopiers and photo-copying apparatus incorporating an optical system.

CONTROL & INSTRUMENTATION

Industrial & Process Control Systems	Automatic regulators or controlling apparatus. Complete systems for data logging, monitoring, displaying, recording and control of processes including telemetry but excluding components or peripherals where separable.
Process Control Instruments	Process measuring and control instruments for temperature, pressure, flow and level. Potentiometric recorders and controllers, panel mounting.
Oscilloscopes	Ray oscilloscopes and oscillographs of the light beam and liquid jet types.
Analytical Instruments	Apparatus for the analysis of gas, liquids and solids including spectrophotometers, spectrographs, spectrofluorimeters, spectropolarimeters and X-ray diffraction/fluorescence apparatus.
Nucleonic Instruments	Instruments and apparatus for measuring or detecting ionising radiations including reactor instruments, radiation measuring and detecting instruments, nucleonic industrial process measuring and control instruments.
Signal Generators	Signal and waveform generators.
Telecommunications Instruments	Cross talk meters, gain measuring instruments, distortion factor meters, psophometers etc.
Machine & Materials Test Instruments	Echo sounding and ultrasonic sounding or detection instruments and apparatus. Inspection and measuring instruments for engineering metrology (surface, texture etc). Instruments and apparatus for balancing mechanical parts.

Electrical Quantity Measuring Instruments	Ammeters, voltmeters, ohmmeters (inc. multimeters) analogue & digital.
Other Test & Measuring Instruments	Other measuring, checking, analysing and controlling instruments and apparatus including apparatus for measuring optical and acoustical quantities Exposure meters, thermocalorimeters and other measuring instruments used in photography and cinematography. Potentiometers and bridges Graph recorders and other recorders not elsewhere specified. Instrumentation tape recorders
Accessories and Parts for Control & Instrumentation	Accessories, parts and spare parts for control and instrumentation.

MEDICAL & INDUSTRIAL

X-Ray & Medical Equipment

Medical X-Ray Equipment	Apparatus based on the use of X-Rays for diagnostic and therapeutic medical and dental purposes, including scanners.
Industrial X-Ray Equipment	Apparatus based on the use of X-Rays for other than medical or dental purposes.
Medical Radiation Equipment	Medical apparatus based on the use of radiation from radioactive substances.
Industrial Radiation Equipment	Apparatus based on the use of radiation from radioactive substances for other than medical use. Electrocardiographs
Other Electromedical Equipment	Patient monitoring equipment Electromedical therapeutic apparatus Electromedical diagnostic apparatus Diathermic apparatus Infra-red and ultraviolet ray apparatus Electroencephalographs Pacemakers, for stimulating heart muscles and other similar apparatus Other electromedical apparatus other than radiological
Hearing Aids	Hearing aids with transistor amplifications.

Industrial Equipment	
Railway Signalling Equipment	Electronic and electrical traffic control and signalling equipment for railways.
Other Traffic Signalling Equipment	Electronic and electrical traffic control equipment for roads, inland waterways and port installations.
Security & Fire Alarms	Electronic and electrical, burglar, fire and similar security alarms and related equipment.
Other Signalling Equipment	Other electronic and electrical sound or visual signalling apparatus such as bells, sirens, buzzers, indicator panels including motor vehicle and civil aircraft signalling equipment.
Induction and Dielectric Equipment	Induction and dielectric heating equipment including furnaces and microwave catering ovens.

COMMUNICATIONS & RADAR

Radar	Radar apparatus for military, civil and meteorology.
Navigational Aids	Ground radio navigational aids Electronic meteorological, hydrological and geophysical instruments and apparatus. Shipborne radio navigational aids. Electronic marine or river navigational instruments and apparatus Airborne radio navigational aids, inc. radio navigational receivers, radio altimeters, inertia navigation systems and ground proximity warning systems. Electronic aeronautical or space navigational instruments and apparatus.
Radio Communications Transmission Apparatus	For radio-telephony and radio telegraphy for ground, shipborne, airborne including those for use in civic aircraft
Radio Communications Reception Apparatus	For radio-telephony and radio telegraphy for ground, shipborne, airborne including those for use in civic aircraft
Transceivers (combined Transmitter-Receivers)	Ground strategic and tactical communications units, two-way mobile radios, handheld and citizen band types, shipborne, airborne including those for use in civil aircraft.
Mobile Radio Telephones	Cellular and other mobile radio telephones

Pocket Pagers	Receivers for paging or calling
Public Broadcasting Transmitters	Radio and television transmitters for public broadcasting
Other Public Broadcast	Radio and television relay links. Television cameras for professional use (domestic video cameras and camera recorders are listed under consumer sector). Closed circuit television.
Other Communications and Military Equipment	Other transmitters, transceivers and receivers not elsewhere specified, including radio remote control apparatus and other airborne electronic equipment.
Accessories and Parts	Accessories, parts and spare parts for communications apparatus and equipment including other equipment not elsewhere specified.

TELECOMMUNICATIONS

Switching Equipment	Telephonic and telegraphic switching apparatus including telephone exchange equipment.
Facsimile Equipment	Facsimile machines
Data and Text Terminal Equipment	Teleprinters, telex machines and other equipment for telegraphic and data communications purposes including modems and multiplexors except those used for computer links only, front end processors, translators etc.
Transmission Equipment	Carrier systems, line apparatus for long distance communications (carrier, duplex, digital multiplexors and repeaters) speech input equipment, telegraph and telephone, radio relay equipment, microwave systems, apparatus for signal conditioning of satellites etc.
Telephone Sets	Telephone handsets, cordless telephones incl. handset and base unit, one and two piece telephones, (mobile radio telephones are included in Communications section)
Other Telecommunications Equipment	Other telecommunications equipment not elsewhere specified inc. headphones and telephone receivers which incorporate loudspeakers, intercoms, telephone answering machines, video teleconferencing, voice mail equipment etc.

Accessories & Parts	Accessories, parts and spare parts for telecommunications apparatus and equipment.

CONSUMER

Consumer Video	
Colour Television	All sizes including colour receivers incorporating radio broadcast receivers and colour receivers incorporating teletext and/or viewdata or similar information systems, video monitors, television projection equipment, (does not include combined TV receivers and VCR)
Monochrome Television	All sizes including monochrome receivers incorporating radio broadcast receivers.
Video Recorders	Television image and sound recorders and reproducers and video disc players, combined television receivers and video recorders.
Video Cameras and Camera Combinations	Cameras for use with domestic video recorders, camera recorder combinations in one case (camcorders).
Consumer Audio	
Portable Radios & Radio Recorders	Battery operated radios and portable radio recorders with or without a clock, AM and FM, stereo and mono types.
Mains Radios & Combination	Clock Radios, mains table radios, radio/record players, radio/tape players, music centres inc. tuners and tuner amplifiers and all kit forms for subsequent assembly. Where a compact disc player constitutes the major part of the combination these are included under CD players.
Car Radios & Combinations	Car radios and car radio/tape player combinations, AM and FM stereo and mono types.
Tape Recorders & Decks	Tape recorders including portable cassette, and open reel recorders/reproducers, cassette and open reel decks. Telephone answering machines are not included (see Telecommunications section)
Record Players & Decks	Record players with/without automatic changing mechanisms including record decks.

Compact Disc Players	Players with laser optical reading system, radio broadcast receivers and car radios which include a CD player; portable CD players.
Consumer Personal	
Electric/Electronic Watches	Pocket watches, wrist watches and other electric/electronic watches.
Electric/Electronic Clocks	Clocks with watch movements including those with a piezo-electric quartz crystal regulation device. Electric clock systems. Battery operated alarm, wall clocks, other mains operated.
Electronic Flashlights	Discharge lamp (electronic) flashlight apparatus

COMPONENTS

Active Components	
Colour Television Tubes	All sized colour television picture cathode ray tubes.
Monochrome Television Tubes	All sized monochrome television picture cathode ray tubes.
X-Ray Tubes	Electronic tubes for use in X-Ray equipment
Other Valves & Tubes	Television camera tubes Image converters or intensifiers Other photocathode tubes Magnetrons Klystrons Travelling wave tubes Other microwave tubes Receiving amplifier valves & tubes Vacuum fluorescent display tubes Other display tubes All other valves & tubes
Diodes	Signal diodes Power rectifiers Voltage reference and regulation diodes Microwave and photosensitive Light emitting diodes Germanium and silicon inc. small signal, power and FETs (field effect transistors).
Thyristors	Thyristors, diacs and triacs

Photoelectric Cells	Photocells
Piezoelectric Crystals	Mounted piezoelectric crystals
Other Discrete Semiconductors	Other discrete semiconductors not elsewhere specified (including photocells and piezoelectric crystals where not specified elsewhere)
Integrated Circuits Linear	Amplifiers, regulators, interfaces, other analogue; including chips but not wafers
Integrated Circuits Digital	Memories, microprocessors, logic circuits, other digital, including chips but not wafers
Integrated Circuits Hybrids	Hybrid integrated circuits
Other Integrated Circuits	Other integrated circuit and microcircuits.
Passive Components	
Electrolytic Capacitors	Aluminium and tantalum fixed capacitors
Other Fixed Capacitors	Ceramic single layer, ceramic multilayer, paper, plastic, other fixed capacitors; power capacitors
Variable Capacitors	Variable and pre-set capacitors
Fixed Resistors	Fixed carbon, composition, film and other excluding heating resistors
Variable Resistors	Variable resistors, rheostats and potentiometers
Multi-Pin and RF Connectors	PCB Connectors, coaxial connectors, rack & panel connectors, other multipin connectors
Other Connection Devices	Plugs and sockets Other electrical apparatus for making connections to or in electrical circuits for a voltage not exceeding 1000 V.
Small Transformers, Chokes, Coils & Other Inductors	Transformers not exceeding 1 kVA Inductors excluding ballasts for discharge tubes.
Relays	Relays
Switches	Isolating and make and break switches Other Switches

Printed Circuits Boards	Printed circuits boards consisting only of conductor elements and contacts Printed circuits boards with other passive elements (PCBs with components, i.e. stuffed boards, are excluded)
Audio Components	
Microphones	Microphones and stands for use with domestic hi-fi systems and professional applications.
Loudspeakers	Loudspeaker for use with domestic and motor vehicles, and for professional applications.
Amplifiers	Audio frequency amplifiers, sound amplifier sets for use with domestic hi-fi and professional applications.
Aerials	Telescopic and whip aerials for portable apparatus and for apparatus fitted in motor vehicles.
	Outside and inside aerials for radio or television broadcast receivers including built-in types and reception via satellite Other aerials.
Unrecorded Media	Prepared media for sound or similar recording but not recorded:
	Magnetic tapes audio, video, cassette and other computer/instrumentation tape
Magnetic film	Magnetic discs, rigid and other Other unprepared media
Cabinets for Radios, TV & Professional Communications Equipment	Cabinets of wood or other materials.
Accessories & Parts for Consumer Equipment	Accessories, parts and spare parts inc. parts for: Microphones, loudspeakers and amplifiers Domestic television receivers and terminal units Domestic and portable radio receivers (tuners etc) Record players, tape recorders and video recorders including sound heads and parts, needles (all types) whether or not mounted and parts of base metal Other

Appendix 5.5
TRANSLATION OF PRODUCT HEADINGS

ENGLISH	FRENCH	GERMAN	ITALIAN
Accessories, Parts and Spare Parts	Accessoires, Parties et Pieces Detachees	Zubehor, Teile und Ersatzteile	Accessori, Parti e Pezzi Staccati
Active Components	Composants Actifs	Aktive Bauelemente	Componenti Attivi
Aerials	Antennes	Antennen	Antenne
Agriculture, Forestry and Fishing	Agriculture, Sylviculture et Peche	Landwirtschaft, Forstwirschaft, und Fischerei	Agricolura Selvicoltura e Pesca
Amplifiers	Amplificateurs	Verstarker	Amplificatori
Analytical Instruments	Instruments d'Analyse	Analysengrate	Strumenti d'Analisi
Area	Superficie	Oberflache	Area
Audio Components	Pieces Detachees pour Appareils de Reproduction du Son	Phonozubehor	Componenti Sono
Cabinets	Meubles	Gehause	Mobili
Cars and Commercial Vehicles Production	Production de Voitures	Kraftfahrzeug Produktion	Produzione de Veicoli
Car Radios	Autoradios	Autoradiogerate	Autoradio
Cassette Tape Recorders	Magnetophone a Cassette	Kassetten-Bandgerate	Registrator a Musicassetta
Cathode Ray Tubes	Tubes Cathodiques	Kathodenstrahlrohren	Tubi Catodici
Chokes and Coils	Bobines d'Inductance	Spulen	Induttanze di Arresto e Bobine
Closed Circuit Television (CCTV)	Television en Circuit Ferme	Kabelfernsehen	Impianti di Televisione a Circuito Chiuso
Colour Television	Recepteurs de Television en Couleur	Farbfernsehempfanger	Televisori a Colori
Colour Television Tubes	Tubes Cathodiques pour Recepteurs de Television en Couleur	Farbbildroehren	Tubi Catodici per Televisori a Colori
Communications	Communication	Drahtlose Nachrichtentechnik	Communicazioni
Compact Disc Players	Tourne-Disques a Systeme de Lecture Optique par Faisceau Laser	Plattenspieler mit Optische Laserstrahl	Compact Disc Giradischi
Components	Composants Electroniques	Beuelemente	Componenti Elettronici
Computers	Ordinateurs	EDV-Anlagen	Calolatori
Computer Systems	Ordinateurs et Systemes	Computer Systeme und Zentraleinheiten	Calcolatori Elettronici e Sistemi
Connectors	Connecteurs	Stechverbinder	Serrafili
Consumer Equipment	Radio-TV Equipment	Fernsehe-Radio-und Phonogerate	Radio e Television Domestica Apparechi
Consumer Prices	Prix de Consommation	Verbraucherpreise	Prezzi al Consumo
Constant Market Prices	Prix a Marche Constant	Konstante Markepreise	Prezzi di Mercato Constante
Control Equipment	Equipement de Controle	Regel-und Steurergerate	Apparecchi di Controllo
Current Market Prices	Prix a Marche Courant	Jeweilige Marktpreise	Prezzi di Mercato Correnti
Data & Text Terminal Equipment	Appareils pour Belinogrammes ou Telephotographie	Gerate fur Telegraphentecknik	Apparecchi per la Telegraphia
Dictation Equipment	Machines a Dicter	Diktiergerate	Macchine per Dettatura
Diodes	Diodes	Dioden	Diodi
Discrete Semiconductors	Semiconducteurs Discrets	Diskrete Halbleiter	Semiconduttori Discreti
Dishwashers	Lave Vaisselle	Geschirrspulmaschinen	Macchine Lavapiatti
Domestic Appliances	Equipement Electromenager	Haushaltsgerate	Electtrodomestici
Electrical Instruments	Instruments Electriques	Elektrische Gerate	Strumenti Elettrici
Electrical Measuring Instruments	Instruments de Mesure Electriques	Elektrische Messgerate	Strumenti de Misura Electici
Electric Cookers	Cuisinieres Electriques	Elektroherde	Cucine Elettriche
Electric Typewriters	Machines a Ecrire	Elektrische	Macchine da Scrivere

ENGLISH	FRENCH	GERMAN	ITALIAN
	Electriques	Schreibmaschinen	Elettriche
Electrocardiographs	Electrocardiographes	Elecktrokardiographen	Elettrocardiografi
Electrolytic Apparatus	Appareils D'Electrolyse	Apparate Gerate fur Elektrolyse	Apparecchi per Elettrolisi
Electronic Capacitor	Condensateurs Electrolytiques	Elektrolyt-Kondensatoren	Condensatori Elettrolitici
Electromedical Equipment	Appareils Electromedicaux	Elekromedizinische Gerate	Apparecchi Elettromedicali
Electronic Accounting Machines	Systemes de Comptabilite Electroniques	Elektronische Buchungsmaschinen	Sistemi Contabili Elettronici
Electronic Calculators	Calculateurs Electroniques	Elektronische Rechenmaschinen	Calcolatrici Elettroniche
Electronic Cash Registers	Caisses Enregistreuses Electroniques	Elektronische Registrierkassen	Registratori di Cassa Elettronici
Electronic Clocks	Pendules Electroniques	Elektronische Uhren	Pendola Elettronici
Electronic Control	Controle Industriel Electronique	Elektronische Steuerung	Controllo Industriale Elettronico
Electronic Data Processing (EDP)	Informatique	Electr. Datenverarbeitung (EDV)	Informatica ed Elaborazione Dati
Electronic Flashlights	Flash Electronique	Elektronische Blitzlichtferate	Flash Elettronici
Electronic Instruments	Instruments Electroniques	Elektronische Instrumente	Strumenti Elettronici
Electronic Musical Instruments	Instruments de Musique Electroniques	Elektronische Musikinstrumente	Strumenti Musicali Elettronici
Electronic Thermometers	Thermometres Electroniques	Elektronische Thermometer	Termometri Elettronici
Electric Thermostats	Thermostats Electroniques	Elektronische Thermostate	Termostati Elettronici
Electronic Watches	Montres Electroniques	Elektronische Uhren	Orologio Elettronici
Employment	Emploi	Erwerbstatige	Occupati
Exchange Rates	Taux de Change	Wechselkurse	Tassi di Cambio
Exchange Rate against the Dollar	Taux de Change par Rapport au Dollar	Dollar Kurse	Tassa di Cambio Dollar
Exports	Exportations	Ausfuhr	Esportazioni
Exports of Goods and Services	Exportations de Biens et Services	Ausfuhr von Waren und Dienstleistungen	Esportazioni di Beni e Servizi
Fixed Capacitors	Condensateurs Fixes	Festkondensatoren	Condensatori Fissi
Fixed Resistors	Resistances Fixes	Festwiderstande	Resistori Fissi
Food Mixers	Batteurs-Melangeurs	Mixgerate	Frullatori
Freezers	Congelateurs	Tiefkuhlgerate	Congelatori
Government Consumption	Consommation Publique	Staatsverbrauch	Consumo Pubblico
Gross Domestic Product	Produit Interieur Brut	Bruttoinlandspodukt	Prodotto Interno Lordo
Gross Fixed Capital Formation	Formation Brute de Capital Fixe	Bruttoanlagein-vestitionen	Investmenti Fissi Lordi
Growth Rate	Taux d'Accroissement	Zuwachsrate	Tassi d'Incremento
Headphones	Casques/Ecouteurs	Kopfhorer	Cuffie
Hearing Aids	Prothese Auditive	Horgerate	Audifoni
Hybrid Circuits	Circuits Hybrides	Hybride Schaltungen	Circuiti Ibridi
Imports	Importations	Einfuhr	Importazioni
Imports of Goods and Services	Importations de Biens Services	Einfuhr von Waren und Dienscleistungen	Importazioni di Beni e Servizi
Indices	Indices	Indices	Indici
Induction and Dielectric Furnaces	Fours d'Induction et Dielectrique	Induction - und Dielektrischofen	Forni ad Induzione ed ad Dielettrice
Inductors	Inducteurs	Inductive Bauelemente	Induttori
Industrial Equipment	Appareils Industriel	Industrielle Gerate	Apparecchi Industriale
Industrial and Process Control Equipment	Appareils de Controle Industriel et de Processus	Industrielle Steuer - und Regeliegerate	Controllo Industriale e di Processi
Industrial Production	Production Industrielle	Industrielle Producktion	Produzione Industriale
Industry	Industrie	Industrie	Industria
Instrumentation	Instrumentation	Messtechnik	Strumentazione
Integrated Circuits	Circuit Integres	Integrierte Schaltungen	Circuiti Intergrati
Lasers	Lasers	Lasers	Lasers
Loudspeakers	Haut-Parleurs	Lautsprecher	Altoparlanti
Machine Tool Control	Equipement de Controle	Werkzeug-Maschinen	Apparecchi di Controllo

ENGLISH	FRENCH	GERMAN	ITALIAN
Equipment	pour Machines-Outils	Steuerungen	per Macchine Utensili
Mains Radio	Recepteurs de Radio Fonctionnant sur source d'energie exterieure	Heimradiogerate	Apparecchi Radio dei Tavolo
Materials Test Instruments	Instruments d'Essai des Materiaux	Materialprufung Gerate	Strumenti di Prova Materiale
Microcircuits	Microcircuits	Mikroschaltungen	Microcircuiti
Microphones	Microphones	Mikrophone	Micronfoni
Millions	Millions	Millionen	Millioni
Mixed Cookers	Cuisinieres Combinees	Kombinierte Herde	Misti Cuicine
Monochrome Television	Recepteurs de Television Monochrome	Schwarzweiss-Fernseh Empfangeer	Televisori Monocromi
Monochrome Television Tubes	Tubes Cathodiques pour Recepteurs de Television Noir et Blanc	Schwarzweiss Empfangeer	Tubi Catodici per Usi Interni
Monolithic Digital	Monolithiques Digitaux	Monolithisch Digital	Monolitici Digitalu
Monolithic Linear	Monolithiques Analogiques (Lineaires)	Monolithisch Analog	Monolitici Analogici
National Currency	Monnaie Nationale	Landswahrung	Moneta Nazionale
National Expenditure	Depense Nationale	Staatlich Verausgabung	Risorse Impiegate per usi
Navigational Aids	Instruments de Navigation	Navigationsgerate	Strumenti per Navigazione
Negligible	Negligeable	Vernachlassigbar	Trascurabile
Not Available	Resultat Inconnu	Nicht Bekannt	Non Conosciuto
Nucleonic Instruments	Instruments de Technique Nucleaire	Strahlungmessgerate	Strumenti Nucleari
Number	Pieces	Stuck	Pezzi
Office Equipment	Machines de Bureau	Buromaschinen	Macchinaria d'Ufficio
Off-Line Equipment	Perforatrices Verificatrices et Calculatrices	Locher Lochpruefer Rechenlocher et Datenerfassungsum Wandlungsgerate	Perforatrici, Verificatrici et Calcolatrici
Optical Instruments	Appareils d'Optiques	Optische Gerate	Apparecchi di Ottico
Oscilloscopes	Oscilloscopes	Oszilographen	Oscilloscopi
Passive Components	Composants Passifs	Passive Bauelemente	Componenti Passivi
Per Capita	Chiffres per Habitant	Je Einwohner	Dati Pro-Capite
Percentage	Pourcentage	Prozentsatz	Percentuale
Peripherals	Equipment Peripherique	Peripheriegerate	Unita Periferiche
Photocopiers	Appareils de Photocopie	Kopierapparat	Apparecchi di Fotocopia
Photoelectric Cells	Cellules Photo-electriques	Photozellen	Cellule Fotoelettriche
Photographic and Cinematographic Instruments	Appareils Photographique et Cinematographique	Photographische und kinematographische Gerate	Apparecchi di Fotografico ed Cinematigrafico
Piezoelectric Crystals	Cristaux Piezoelectriques	Piezoelektrische Kristalle	Cristalli Piezoelettrici
Pocket Pagers	Recepteurs de Poche	Taschenempfger fur	Apparecchi Receventi Personenrufanlag Tascabili
Population	Population Totale	Gesamtbevolkerung	Popolazione Totale
Portable Radios	Recepteurs de Radio Portatifs	Taschen u Kofferadiogerate	Apparecchi Radio Portatili
Printed Circuits	Circuits Imprimes	Gedruckte Schaltungen	Circuit Stampati
Private Consumption	Consommation Privee	Privater Verbrauch	Consumi Privati
Process Control Instruments	Instruments de Control de Processus	Prozess-Steuerungs-gerate	Strumenti di Controllo di Processi
Public Broadcasting	Materiel de Radiodiffusion et Television	Rundfunk	Apparecchi Transmettenti per Radio e TV
Radar (Ground)	Radars (au Sol)	Radargerate (am Boden)	Radar (Terra)
Radar (Shipborne)	Radars (Embarques)	Radargerate (fur Wasserfahrzeuge)	Radar (Navali)
Radar (Airborne)	Radars (Aeroportes)	Radargerate (fur Luftfahrzeuge)	Radar (Aerei)
Radiation Equipment	Appareils de radiation	Strahlungs Gerate	Apparecchi di Radiazioni
Radio Combinations	Combine Radio-Phono/ Magnetophone	Musiktruhen	Apparecchi Radio Combinati per Giradischi e/o Giranastri
Radio Communications	Materiel Radioelectrique	Funkgerate	Radio Professionale
Radio Recorders	Combine Radio-Magnetophone	Radio Recorder	Radio Registratori
Radio and TV Tuners	Tuners Radio et TV	Tuner fur Rundfunk und Fernsehgerate	Sintonizzatori per Radio e TV

ENGLISH	**FRENCH**	**GERMAN**	**ITALIAN**
Railway Signalling Equipment	Appareils de Signalisation Pour Chemins de Fer	Signalgerate fur Eisenhahnen	Apparecchi di Segnalamento Per Ferovie
Receivers	Recepteurs	Empfangsgerate	Receviotri
Record Players	Tourne-Disques	Plattenspieler	Giradischi
Refrigerators	Refrigerateurs	Kuhlschranke	Frigoriferi
Relays	Relais	Relais	Rele
Seasonally Adjusted	Adjuste Periodiquement	Saisonbereinigte	Aggiustamento Stagionale
Security and Fire Alarms	Systeme de Securite et d'Alarme au Feu	Sicherheit du und Feueralarm	Allarme di Securezza e di Fuoco
Semiconductors	Semiconducteurs	Halbleiter Beuelemente	Semiconduttori
Signal Generators	Generateurs De Signaux	Prufgeneratoren	Generatori di Segnali
Signalling Equipment	Appareils de Signalisation	Signalgerate	Apparecchi di Segnalemento
Small Transformers	Petits Transformateurs	Kleine Transformatoren	Transformatori Piccoli
Standard of Living	Niveau de Vie	Lebensstandard	Standard di Vita
Switches	Interrupteurs	Schalter	Interruttori
Switching Equipment	Materiel de Commutation	Telexvermittlungs-einrichtungen	Apparacchi Telescriventi
Tape Decks	Plantines de Magnetophone	Tonbanddecks	Piastre a Nastro
Tape Players	Lecteurs Magnetiques	Magnetband - Wiedergabegerate	Riproduttori di Nastri
Tape Recorders	Magnetophones	Tonbandgerate	Magnetofoni
Telecommunications	Telecommunications	Fernmeldetechnik	Telecomunicazioni
Telephone Sets	Postes Telephoniques	Telefonapparate	Apparecchi Telefonici
Television Cameras	Camera de Television	Fernsehapparat - Kamera	Macchina Fotographica di Televisione
Test and Measuring Instruments	Equipements D'Essai et de Mesure	Pruf - und Mess Gerate	Strumenti di Misura e di Prova
Thyristors	Thyristors	Thyristoren	Tiristori
Traffic Signalling Equipment	Appareils de Signalisation pour la Circulation	Signalgerate fur Verkehr	Apparecchi di Segnalamento per Traffico Stradale
Transceivers	Recepteurs et Emetteurs Reunis	Empfangsgerate und Rundfunksen-Deanlagen Vereinet	Combinazione di Recevitori ed Trasmettitori
Transistors	Transistors	Transistoren	Transistori
Transmission Equipment	Equipement de Transmission	Urbertragung-seinrichtungen	Apparecchi di Transmissione
Transmitters	Emetteurs	Rundfunksendeanlagen	Trasmettitori
Tumble Dryers	Sechoirs a Tambours	Waschetrockner	Asciugatrici a Tamburo
Turntable Units	Tourne-Disques	Plattenlaufwerk	Giradischi
TV Games	Jeux Video	Video Spieler	Video Gioco
Ultrasonic Instruments	Instruments de Mesure a Ultrasons	Ultraschallmessgerate	Strumenti di Misura Ultrasonici
Unrecorded Media	Bandes Magnetiques et Disques Vierges	Unbespielte Magnetton-bander un Platten	Nastri e Dischi Magnetici Vergini
Vacuum Cleaners	Aspirateurs	Staubsauger	Aspiravolvere
Valves and Tubes	Tubes Electroniques	Elektronenrohren	Valvole e Tubi
Variable Capacitors	Condensateurs Variable	Einstellbare Kondensatoren	Condensatori Variabili
Variable Resistors	Resistances Variables	Einstellbare Widerstande	Resistori Variabili
Video Cameras	Appareils de Vues pour Television	Video Kameras	Video Macchina Fotografica
Video Recorders	Magnetoscope	Video Recorder	Video Registratori
Wages	Salaires	Verdienste	Salari
Washing Machines	Machines a Laver	Waschmaschinen	Lavabiancheria
Watches and Clocks Production	Production de Montres et Horloges	Uhrenproduktion	Produzione de Sveglie ed Orlogi
Wholesale Prices	Prix de Gros	Grosshandelspreis	Prezzio All'ingrosso
X-Ray Equipment	Equipment a Rayons X	Rontgengerate	Apparecchiature per Raggi X

Printed in Dunstable, United Kingdom

91221465R00107